Regulation and Effect of Taurine on Metabolism

Regulation and Effect of Taurine on Metabolism

Editors

Teruo Miyazaki
Takashi Ito
Alessia Baseggio Conrado
Shigeru Murakami

MDPI • Basel • Beijing • Wuhan • Barcelona • Belgrade • Manchester • Tokyo • Cluj • Tianjin

Editors
Teruo Miyazaki
Joint Research Center
Tokyo Medical University
Ibaraki Medical Center
Ami
Japan

Takashi Ito
Department of Bioscience and Biotechnology
Fukui Prefectural University
Eiheiji
Japan

Alessia Baseggio Conrado
National Heart and Lung Institute
Imperial College London
London
United Kingdom

Shigeru Murakami
Department of Bioscience and Biotechnology
Fukui Prefectural University
Eiheiji
Japan

Editorial Office
MDPI
St. Alban-Anlage 66
4052 Basel, Switzerland

This is a reprint of articles from the Special Issue published online in the open access journal *Metabolites* (ISSN 2218-1989) (available at: www.mdpi.com/journal/metabolites/special_issues/Taurine_Metabolism).

For citation purposes, cite each article independently as indicated on the article page online and as indicated below:

LastName, A.A.; LastName, B.B.; LastName, C.C. Article Title. *Journal Name* **Year**, *Volume Number*, Page Range.

ISBN 978-3-0365-6875-1 (Hbk)
ISBN 978-3-0365-6874-4 (PDF)

Contents

About the Editors

Teruo Miyazaki

Teruo Miyazaki is an associate professor of the Joint Research Center at Tokyo Medical University Ibaraki Medical Center in Japan. His main research interests are in the organ–organ interaction of metabolism of taurine, amino acids, and energy production among the skeletal muscle, liver, intestine, and brain in exercise and metabolic diseases. He currently serves as an editorial board member of the nutrition and metabolism section in the journal of *Metabolites* in MDPI and a board member of the Japan Taurine Society.

Takashi Ito

Takashi Ito is a professor of Fukui Prefectural University, where he teaches bioscience regarding animals and human health and he researches the beneficial effects of food chemicals. He has an expertise in the physiological role of taurine in peripheral tissues, including the heart, muscle, and liver. He has investigated the influence of taurine deficiency in mouse model, such as taurine transporter knockout mouse. In some research articles, his group illustrated that taurine transporter knockout mice display cardiomyopathy, skeletal muscle atrophy, exercise intolerance, premature aging, and other physiological disorders. His current research interest includes the pharmacological effects of taurine and its derivatives. They recently reported the effect of taurine and N-chlorotaurine against lipopolysaccharide-induced lung inflammation in mice; a part of these investigations has been published in this Special Issue, "N-Chlorotaurine Reduces the Lung and Systemic Inflammation in LPS-Induced Pneumonia in High Fat Diet-Induced Obese Mice ". He currently serves as a board member of the Japan Taurine Society.

Alessia Baseggio Conrado

Alessia Baseggio Conrado is a clinical project manager of the Children's Clinical Research Facility's at St Mary's Hospital within the Imperial College London. Her main research interests are in oxidative stress, sulfur compound, and their protective role, phototoxicity induced by drugs and chemical compounds, investigation on the allergic reactions and the mechanisms of food-triggered anaphylaxis.

Shigeru Murakami

Shigeru Murakami is a professor of Fukui Prefectural University. He specializes in the functional evaluation of food ingredients and pharmaceuticals, particularly targeting their actions in metabolic diseases, such as obesity, diabetes, and dyslipidemia. In his taurine research, he mainly uses cultured cells and laboratory animals to elucidate the physiological and pharmacological actions of taurine. He has revealed the role of taurine in preventing metabolic diseases and maintaining skin function, as well as the mechanism of action of taurine. He currently serves as president of the Japan Taurine Society and as a board member of the International Taurine Society.

Preface to "Regulation and Effect of Taurine on Metabolism"

In this Special Issue "Regulation and Effect of Taurine on Metabolism", the newest insight into taurine (2-aminoethanesulfonic acid) role has been reported from the scientific community which investigates its wide-ranging physiological and pharmacological play. Taurine could have been erroneously considered a small and simple amino acid but since its discovery, almost 200 years ago, has proven to play a crucial role as an essential compound in numerous fields, including biochemistry, physiology, biology, molecular biology, basic and clinical medicine, sport and exercise science, neuroscience, radiology, immunology, nutrition, food science, marine science, and others.

The aim of this Special Issue was to investigate taurine key nutrient properties associated with various metabolisms to maintain healthy conditions, prevent various diseases, and support growth and aging.

Thanks to the authors'contributions, new understandings were achieved on the fetuses and infants'growth of cells and tissue development; on its function in the brain, skeletal muscles, testis, and on body temperature; as well as its role to prevent and improve metabolic diseases as obesity, diabetes, and fatty liver. Moreover, possible regulatory protective roles were also reported against metabolic diseases including diabetes, hypertension, and inflammatory diseases, and its potential support in the treatment of Duchenne muscular dystrophy.

The editors hope this Special Issue will promote and facilitate further investigations on the clinical importance of taurine and support the taurine scientific community in disseminating their research findings.

Teruo Miyazaki, Takashi Ito, Alessia Baseggio Conrado, and Shigeru Murakami
Editors

MDPI

Editorial

Editorial for Special Issue on "Regulation and Effect of Taurine on Metabolism"

Teruo Miyazaki [1,*], Takashi Ito [2], Alessia Baseggio Conrado [3] and Shigeru Murakami [2]

1 Joint Research Center, Tokyo Medical University Ibaraki Medical Center, Ami 300-0395, Ibaraki, Japan
2 Faculty of Biotechnology, Fukui Prefectural University, Eiheiji 910-1195, Fukui, Japan
3 National Heart and Lung Institute, Imperial College London, Norfolk Place, London W2 1PG, UK
* Correspondence: teruom@tokyo-med.ac.jp

Taurine (2-aminoethanesulfonic acid) is well known to be abundantly contained in almost all the tissues and cells of various mammals, fish, and shellfish [1]. This abundance of taurine is maintained in two major ways. One is endogenous biosynthesis from sulfur-containing amino acids, and the methionine-cysteine metabolic pathway, which mainly occurs in the liver and some other organs including the kidneys, brain, and adipose tissue [2,3]. The second is an exogeneous mechanism involving a specific transporter, the taurine transporter (TAUT/SLC6a6), which facilitates the intestinal absorption of taurine from dietary animal protein and cellular uptake from circulating blood [4]. Taurine has been reported to play wide-ranging physiological and pharmacological roles, from being involved in basic cell processes such as differentiation, growth, and aging [5,6] to being implicated in the prevention and therapy of liver and heart diseases and inherited metabolic disorders, as well as the benefits of exercise in healthy individuals [6]. In the Special Issue entitled *Regulation and Effect of Taurine on Metabolism* published in the *Metabolites* journal, eight original and two review articles introduce the newest findings of taurine on certain metabolism in humans and animals such as fish.

Taurine is well-known to be an essential nutrient for fetuses and infants during the development and growth of cells and tissues. The ability to biosynthesize taurine is low in the first period of life, and therefore, external intake is essential. Tochitani reviewed the importance of taurine transfer via the placenta and breast milk from the mothers to the fetuses and infants of mammalians, focusing on regulatory mechanisms [7]. During pregnancy, maternal blood circulating in the placenta is concentrated with taurine, allowing its efficient transfer to the fetus' blood through TAUT present in the placenta. After delivery, taurine is provided to the neonate through breast milk, and its content is the highest during the first days of lactation. In the rat mammary gland, the expression of TAUT and key biosynthetic enzymes is increased during pregnancy and early lactation. The taurine transferred into the fetus and infant are mostly delivered to the brain across the blood–brain barrier (BBB) by TAUT, where it behaves as an inhibitory neurotransmitter—an agonist for γ-aminobutyric acid (GABA) and glycine receptors. Consequently, it promotes the structural and functional maturation of the brain during these periods. In this review, Tochitani also discussed the importance of taurine as a determining factor for health and disease later in life [7].

In addition to its role in the fetal and infant periods, taurine is an important factor for functions in the brain as well as other organs, such as the skeletal muscles, of adults. Watanabe et al. evaluated the influence of taurine depletion on anxiety-like behavior, skeletal muscle function, and body growth using *Taut*-KO mice [8]. In an elevated plus maze test, *Taut*-KO mice were observed to have decreased anxiety-like behavior and difficulties in making decisions underlying an approach-avoidance conflict and risk assessment. The hearing ability was also reduced in *Taut*-KO mice. In addition, a reduction in muscular endurance and lower body growth were observed during the development period from

Citation: Miyazaki, T.; Ito, T.; Baseggio Conrado, A.; Murakami, S. Editorial for Special Issue on "Regulation and Effect of Taurine on Metabolism". *Metabolites* **2022**, *12*, 795. https://doi.org/10.3390/metabo12090795

Received: 24 August 2022
Accepted: 24 August 2022
Published: 26 August 2022

postnatal day 0 until day 60. This suggested that taurine depletion could lead to imbalances in muscular energy production and induce the anorexigenic effect of an adipose-cell-released hormone, leptin, which mediates satiety through the STAT3 pathway. This was confirmed by the phosphorylation of STAT3 being very strongly induced in the brains of the *Taut*-KO mice. Furthermore, Watanabe et al. suggested a possibility that the lower skeletal muscle functions in *Taut*-KO mice could be affected by the activation of STAT3 [8].

Elhussiny et al. evaluated the role of taurine in the brain in the regulation of body temperature and stress behavior in a neonatal chick social isolation stress model induced by the intracerebroventricular (ICV) injection of corticotropin-releasing factor [9]. This factor modulates stress-related effects in the central nervous system [10] and is an indicator of hypothalamic–pituitary–adrenal axis activity [11]. In the neonatal chick model, the ICV injection of taurine alleviated the rectal hyperthermia and stress behaviors such as the number of distress vocalizations and durations of active wakefulness/sleeping postures. In the brains, taurine administration also decreased glutamate, cystathionine, and cysteine, which are substrates and precursors of the antioxidant molecule glutathione. Moreover, it decreased leucine, isoleucine, and alanine, which are related to glutamate synthesis, suggesting that taurine enhanced glutathione synthesis, regulating the stress response through inducing sedative and hypnotic effects.

Similar to the transport of taurine across the BBB by TAUT, taurine was observed to be transported across the blood–testis barrier (BTB) through TAUT, which tightly limits the transport of molecules from the circulating blood into the testis. Taurine is abundant in human semen (where it is 10-fold higher than in the blood) and is suggested to protect the motility and form of sperm from oxidative stress in the testis, where antioxidant enzymes are weakly expressed [12]. Kubo et al. investigated the blood-to-testis transport of taurine using an integration plot analysis in mice and a mouse-derived Sertoli cell line (TM4 cell) [13]. In this analysis, to evaluate the apparent influx clearance, [^{3}H]taurine was injected into the mouse's internal jugular vein and transported into the testis from the circulating blood (approximately 3-fold more than that of a non-permeable paracellular transport marker: [^{14}C]D-mannitol). In TM4 cells, the uptake of [^{3}H]taurine was time- and dose-dependent, and also Na^+- and Cl^--dependent, but was inhibited by other substrates of TAUT such as β-alanine. This study also confirmed the presence of the *Taut* gene and its protein expressions in the mouse testis and TM4 cells as well as the distribution of its protein in the seminiferous tubules of the mouse testis. The findings of Kubo et al. [13] showed that the higher abundance of taurine in semen is due to transport across the BTB by TAUT.

Taurine is reported to prevent and improve metabolic diseases as obesity, diabetes, and fatty liver in animal models and humans [14,15]. In this Special Issue, Murakami et al. [16] and Chen et al. [17] describe the ameliorative effects of oral taurine administration on diabetes and fatty liver in a model mouse, and high-fat-diet-induced fatty liver in fish, respectively. Satsu et al. [18] analyzed the mechanism underlying the enhancing effect of taurine on the transcriptional activity of thioredoxin-interacting protein (TXNIP), whose regulatory effects protect against metabolic diseases including diabetes and hypertension and inflammatory diseases including ulcerative colitis [19,20].

In addition to the evidence for beneficial effects of taurine administration on type 2 diabetes animal models, Murakami et al. evaluated the effect of taurine on streptozotocin (STZ)-induced type 1 diabetic mice by focusing on glucose metabolism and oxidative stress [16]. Chronic oral taurine administration alleviated STZ-induced hyperglycemia and hyperketonemia and maintained hepatic glycogen content, with the upregulation of the hepatic glucose transporter (*Glut2*) gene and suppression of oxidative stress in the liver.

Chen et al. [17] investigated the mechanisms underlying the attenuative effects of dietary taurine administration on fatty liver in groupers (*Epinephelus coioides*). Taurine supplementation significantly inhibited the increases in lipid content and tissue-weight-to-body-weight ratio in the liver induced by a chronic high-fat diet. According to transcriptomic analysis, 160 (72 up- and 88 downregulated) differentially expressed genes

(DEGs) were induced by the high-fat diet, and 49 (26 up- and 23 downregulated) of them were identified as being affected by taurine supplementation. Among them, the enriched DEGs involved the upregulation of the bile secretion pathway, fatty acid β-oxidation, and taurine synthesis, and the downregulation of phospholipase D signaling and glycolysis were suggested to be possible mechanisms of the attenuative effect of dietary taurine on high-fat-diet-induced fatty liver in fish. This evidence supports the idea that taurine could improve the efficiency of the aquaculture of fish using high-fat feeds by protecting them against fatty liver syndrome.

Through reporter assays, Satsu et al. confirmed that taurine increased Est-1 protein's binding to the TXNIP promoter region by the identification of the taurine response element region using human intestinal Caco-2 cells transfected with mutation vectors. Furthermore, taurine was also confirmed to increase the phosphorylation at Thr38 of the Ets-1 protein and activate the ERK1/2 pathway by enhancing phosphorylation; the upregulation of TXNIP gene expression was not observed with p38 or JNK, which also belong to the MAP kinase family. Finally, Satsu et al. suggested that unknown taurine receptors that recognize intracellular/extracellular taurine, leading to ERK–Est-1 activation, might exist in the intestinal cells [18]. On the other hand, the GABA and glycine receptors in the brain are well-known to be taurine receptors [21].

Merck and De Paepe reviewed the role of taurine in skeletal muscle function and its potential support in the treatment of Duchenne muscular dystrophy (DMD) [22]. The authors reviewed what is known about taurine's role in the physiological functions of skeletal muscles, including the impact of taurine depletion in skeletal muscles in *Taut*-KO mice and in mice treated with the TAUT antagonist guanodinoethane sulfonate. Furthermore, they discussed the roles of taurine in osmotic homeostasis, protein and membrane stabilization, oxidative stress, mitochondrial protein synthesis, and calcium homeostasis in skeletal muscle cells. Subsequently, the pathological characteristics of taurine in an X-chromosome-linked muscular dystrophy (mdx) mouse model and DMD patients were introduced, and current findings on the effects of taurine supplementation on the muscle force, oxidative stress, inflammation, E–C coupling, and histopathological characteristics in the mdx mouse model were discussed. In addition, the synergistic effects of taurine treatment with a corticosteroid (α-methylprednisolone) on muscle strength and muscle atrophy in the mdx mouse were reviewed, including the counteractive effects of taurine on the side effects of a corticosteroid (dexamethasone), such as muscle atrophy and mitochondrial dysfunction in bone.

Although taurine is known not to be metabolized as an end-product of the methionine–cysteine metabolic pathway, the amino group of taurine conjugates with a short-chain fatty acid acetate that is a product of alcoholic detoxification in the liver and, consequently, is converted to *N*-acetyltaurine (NAT) [23]. Miyazaki et al. showed, in a human study, that taurine prevents the accumulation of mitochondrial acetyl-CoA metabolized from acetate derived from the liver in the skeletal muscles during endurance exercise, through the production of the taurine derivative NAT, which conjugates with the excess acetate and promotes its excretion from the skeletal muscle into the urine [24].

Another type of taurine derivative, *N*-chlorotaurine (also known as taurine chloramine; TauCl), is formed by the reaction of the amino group of taurine with excess hypochlorous acid produced via a halide-dependent myeloperoxidase system in phagocytic cells, especially neutrophils and macrophages, where the intercellular taurine concentrations reach 50–100 mM (600-fold higher than those in the plasma). TauCl acts as an anti-inflammatory agent through killing a broad spectrum of pathogenic microbes (viruses, fungi, and bacteria) [25] and has a suppressive effect on inflammatory cytokines [26]. Khanh Hoang et al. reported new findings regarding the effects of TauCl treatment on pulmonary and systemic inflammation in lipopolysaccharide (LPS)-induced pneumonia in high-fat-diet-induced obese mice [27]. In this model, TauCl treatment attenuated lung edema, accompanied by the inhibition of the gene expression and serum levels of the proinflammatory cytokines

IL-6 and TNFα. Furthermore, the skeletal muscle wasting associated with LPS-induced pneumonia was alleviated by the TauCl treatment.

In conclusion, the newest insights reported in this Special Issue confirm the many various biological roles of taurine. One of the reasons for its various roles could be its simple and specific structure that is very similar to the structures of other β-amino acids with amino and sulfo groups, and it has actually been proved to be an agonist for neuroreceptors of the γ-amino acid GABA and α-amino acid glycine. Moreover, taurine's properties could be explained considering the amino group itself, which plays a relevant role in the conjugative and/or transfer reactions with metabolites including bile acids, hypochlorous acid, fatty acids, and mitochondrial transfer RNA, and, last but not least, the osmoregulation and detoxification properties, based on the high hydrophilicity of taurine's sulfo group.

Author Contributions: Writing—original draft preparation and project administration: T.M.; writing—review and editing: T.I., A.B.C. and S.M. All authors have read and agreed to the published version of the manuscript.

Funding: This research received no external funding.

Conflicts of Interest: Not applicable.

References

1. Jacobsen, J.G.; Smith, L.H. Biochemistry and physiology of taurine and taurine derivatives. *Physiol. Rev.* **1968**, *48*, 424–491. [CrossRef] [PubMed]
2. Tappaz, M.L. Taurine biosynthetic enzymes and taurine transporter: Molecular identification and regulations. *Neurochem. Res.* **2004**, *29*, 83–96. [CrossRef]
3. Murakami, S. The physiological and pathophysiological roles of taurine in adipose tissue in relation to obesity. *Life Sci.* **2017**, *186*, 80–86. [CrossRef] [PubMed]
4. Baliou, S.; Kyriakopoulos, A.M.; Goulielmaki, M.; Panayiotidis, M.I.; Spandidos, D.A.; Zoumpourlis, V. Significance of taurine transporter (taut) in homeostasis and its layers of regulation (review). *Mol. Med. Rep.* **2020**, *22*, 2163–2173. [CrossRef] [PubMed]
5. Sturman, J.A. Taurine in development. *Physiol. Rev.* **1993**, *73*, 119–147. [CrossRef]
6. Jong, C.J.; Sandal, P.; Schaffer, S.W. The role of taurine in mitochondria health: More than just an antioxidant. *Molecules* **2021**, *26*, 4913. [CrossRef]
7. Tochitani, S. Taurine: A maternally derived nutrient linking mother and offspring. *Metabolites* **2022**, *12*, 228. [CrossRef]
8. Watanabe, M.; Ito, T.; Fukuda, A. Effects of taurine depletion on body weight and mouse behavior during development. *Me-tabolites* **2022**, *12*, 631. [CrossRef]
9. Elhussiny, M.Z.; Tran, P.V.; Tsuru, Y.; Haraguchi, S.; Gilbert, E.R.; Cline, M.A.; Bungo, T.; Furuse, M.; Chowdhury, V.S. Central taurine attenuates hyperthermia and isolation stress behaviors augmented by corticotropin-releasing factor with modifying brain amino acid metabolism in neonatal chicks. *Metabolites* **2022**, *12*, 83. [CrossRef]
10. Owens, M.J.; Nemeroff, C.B. Physiology and pharmacology of corticotropin-releasing factor. *Pharmacol. Rev.* **1991**, *43*, 425–473.
11. Arborelius, L.; Owens, M.J.; Plotsky, P.M.; Nemeroff, C.B. The role of corticotropin-releasing factor in depression and anxiety disorders. *J. Endocrinol.* **1999**, *160*, 1–12. [CrossRef] [PubMed]
12. Holmes, R.P.; Goodman, H.O.; Shihabi, Z.K.; Jarow, J.P. The taurine and hypotaurine content of human semen. *J. Androl.* **1992**, *13*, 289–292. [PubMed]
13. Kubo, Y.; Ishizuka, S.; Ito, T.; Yoneyama, D.; Akanuma, S.I.; Hosoya, K.I. Involvement of TauT/SLC6A6 in taurine transport at the blood-testis barrier. *Metabolites* **2022**, *12*, 66. [CrossRef]
14. Imae, M.; Asano, T.; Murakami, S. Potential role of taurine in the prevention of diabetes and metabolic syndrome. *Amino Acids* **2012**, *46*, 81–88. [CrossRef]
15. Franconi, F.; Loizzo, A.; Ghirlanda, G.; Seghieri, G. Taurine supplementation and diabetes mellitus. *Curr. Opin. Clin. Nutr. Metab. Care* **2006**, *9*, 32–36. [CrossRef] [PubMed]
16. Murakami, S.; Funahashi, K.; Tamagawa, N.; Ning, M.; Ito, T. Taurine ameliorates streptozotocin-induced diabetes by modulating hepatic glucose metabolism and oxidative stress in mice. *Metabolites* **2022**, *12*, 524. [CrossRef]
17. Chen, M.; Bai, F.; Song, T.; Niu, X.; Wang, X.; Wang, K.; Ye, J. Hepatic transcriptome analysis provides new insight into the lipid-reducing effect of dietary taurine in high-fat fed groupers (*Epinephelus coioides*). *Metabolites* **2022**, *12*, 670. [CrossRef]
18. Satsu, H.; Gondo, Y.; Shimanaka, H.; Imae, M.; Murakami, S.; Watari, K.; Wakabayashi, S.; Park, S.J.; Nakai, K.; Shimizu, M. Signaling pathway of taurine-induced upregulation of TXNIP. *Metabolites* **2022**, *12*, 636. [CrossRef]
19. Basnet, R.; Basnet, T.B.; Basnet, B.B.; Khadka, S. Overview on thioredoxin-interacting protein (TXNIP): A potential target for diabetes intervention. *Curr. Drug Targets* **2022**, *23*, 761–767. [CrossRef]
20. Takahashi, Y.; Masuda, H.; Ishii, Y.; Nishida, Y.; Kobayashi, M.; Asai, S. Decreased expression of thioredoxin interacting protein mRNA in inflamed colonic mucosa in patients with ulcerative colitis. *Oncol. Rep.* **2007**, *18*, 531–535. [CrossRef]

21. Oja, S.S.; Saransaari, P. Significance of taurine in the brain. *Adv. Exp. Med. Biol.* **2017**, *975*, 89–94. [CrossRef] [PubMed]
22. Merckx, C.; De Paepe, B. The role of taurine in skeletal muscle functioning and its potential as a supportive treatment for Duchenne muscular dystrophy. *Metabolites* **2022**, *12*, 193. [CrossRef] [PubMed]
23. Shi, X.; Yao, D.; Chen, C. Identification of *N*-acetyltaurine as a novel metabolite of ethanol through metabolomics-guided biochemical analysis. *J. Biol. Chem.* **2012**, *287*, 6336–6349. [CrossRef] [PubMed]
24. Miyazaki, T.; Nakamura-Shinya, Y.; Ebina, K.; Komine, S.; Ra, S.G.; Ishikura, K.; Ohmori, H.; Honda, A. *N*-acetyltaurine and acetylcarnitine production for the mitochondrial acetyl-CoA regulation in skeletal muscles during endurance exercises. *Me-tabolites* **2021**, *11*, 522. [CrossRef]
25. Marcinkiewicz, J.; Walczewska, M. Neutrophils as sentinel cells of the immune system: A role of the MPO-halide-system in innate and adaptive immunity. *Curr. Med. Chem.* **2020**, *27*, 2840–2851. [CrossRef]
26. Kim, C.; Cha, Y.N. Taurine chloramine produced from taurine under inflammation provides anti-inflammatory and cytoprotective effects. *Amino Acids* **2014**, *46*, 89–100. [CrossRef]
27. Khanh Hoang, N.; Maegawa, E.; Murakami, S.; Schaffer, S.W.; Ito, T. *N*-chlorotaurine reduces the lung and systemic inflammation in LPS-induced pneumonia in high fat diet-induced obese mice. *Metabolites* **2022**, *12*, 349. [CrossRef]

 metabolites

MDPI

Review

Taurine: A Maternally Derived Nutrient Linking Mother and Offspring

Shiro Tochitani [1,2,3,4]

[1] Division of Health Science, Graduate School of Health Science, Suzuka University of Medical Science, Suzuka 513-8670, Japan; tochitani.shiro@gmail.com; Tel.: +81-59-373-7069
[2] Department of Radiological Technology, Faculty of Health Science, Suzuka University of Medical Science, Suzuka 513-8670, Japan
[3] Center for Preventive Medical Sciences, Chiba University, Chiba 263-8522, Japan
[4] Department of Neurophysiology, Hamamatsu University School of Medicine, Hamamatsu 431-3192, Japan

Abstract: Mammals can obtain taurine from food and synthesize it from sulfur-containing amino acids. Mammalian fetuses and infants have little ability to synthesize taurine. Therefore, they are dependent on taurine given from mothers either via the placenta or via breast milk. Many lines of evidence demonstrate that maternally derived taurine is essential for offspring development, shaping various traits in adults. Various environmental factors, including maternal obesity, preeclampsia, and undernutrition, can affect the efficacy of taurine transfer via either the placenta or breast milk. Thus, maternally derived taurine during the perinatal period can influence the offspring's development and even determine health and disease later in life. In this review, I will discuss the biological function of taurine during development and the regulatory mechanisms of taurine transport from mother to offspring. I also refer to the possible environmental factors affecting taurine functions in mother-offspring bonding during perinatal periods. The possible functions of taurine as a determinant of gut microbiota and in the context of the Developmental Origins of Health and Disease (DOHaD) hypothesis will also be discussed.

Keywords: 2-aminoethanesulfonic acid; developmental programming; $GABA_A$ receptors; glycine receptors; neural progenitor; neural stem cell; perinatal nutrition; placental transfer; obesity; taurine transporter

Citation: Tochitani, S. Taurine: A Maternally Derived Nutrient Linking Mother and Offspring. *Metabolites* **2022**, *12*, 228. https://doi.org/10.3390/metabo12030228

Academic Editors: Teruo Miyazaki, Takashi Ito, Alessia Baseggio Conrado and Shigeru Murakami

Received: 10 December 2021
Accepted: 3 March 2022
Published: 5 March 2022

1. Introduction

A significant development in the evolution of mammals is placentation, intrauterine development of the fetus, and extensive care after birth that improves infant survival to a reproductive age [1]. Only female mammals have the ability to provide prenatal resources through the placenta and produce milk for postnatal development [2]. Not surprisingly, females form their strongest social bonds with their offspring [1]. In addition, offspring development can be affected by early mother–offspring relationships [3]. Thus, it is essential to understand the nature of mother–offspring bonding during pregnancy and the postpartum period and the factors that affect bonding [3].

Taurine (2-aminoethanesulfonic acid) is a sulfur-containing organic acid with various biological functions, including membrane stabilization, cell volume regulation, mitochondrial protein translocation, anti-oxidative activity, and modulation of intracellular calcium levels [4,5]. In addition, taurine structurally resembles neurotransmitter γ-aminobutyric acid (GABA) and glycine and interacts with both $GABA_A$ and glycine receptors to induce chloride currents in neuronal cells [5–7].

In mammals, adults synthesize taurine in the liver from methionine/cystine, although fetuses and infants have limited ability to synthesize taurine because they have limited levels of γ-cystathionase and cysteine sulfinic acid (CSAD) in livers and brains (Figure 1) [8]. Fetuses and infants depend on the taurine supplied by mothers via the placenta or breast milk [5,9]. Taurine is a principal constituent of the amino acid pool in the milk in many species, including humans, chimpanzees, baboons, rhesus monkeys, Java monkeys, sheep, and rats [10,11]. Taurine has the second highest concentration in breast milk after glutamate in these species [10]. Notably, in cats, in which taurine deficiency during early development leads to severe morphological developments of the retina and cerebellum [12,13], the concentration of taurine in milk is very high (2.87 M), being second to that in gerbil (5.95 M) [8]. Taurine is a molecule that links the mother with the offspring. In this review, I provide a comprehensive overview of the functions of taurine in offspring development, the regulatory mechanisms of taurine transport from mother to offspring, and the outcomes of taurine depletion during development.

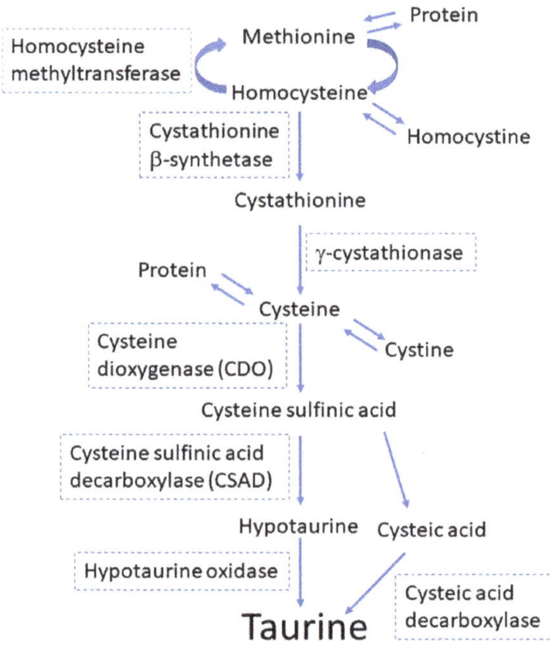

Figure 1. The principal pathway for the synthesis of taurine from methionine and cysteine.

2. Transport of Taurine from Mothers to Offspring

2.1. Cloning of a Taurine Transporter and Its Function

The molecular system for taurine transport has been revealed by conducting various physiological analyses. The taurine transport system is energized by a Na^+ gradient and requires Cl^-, and its activity is inhibited by Ca^{2+}/diacylglycerol-dependent protein kinase (PKC) [14]. In addition, taurine efflux is stimulated by hypoosmotic conditions [15]. Cloning cDNAs for taurine transporter (TauT) revealed an aspect of the molecular basis for taurine transport with these physiological properties. The cDNAs encoding for the taurine transporter (TauT) were first cloned using sequence similarities to glycine, GABA, or other neurotransmitter transporters from Madin–Darby canine kidney cells and mouse and rat brains [16–18]. Then, human TauT was cloned from the human placenta using a similar approach [19]. The Na^+/Cl^-/taurine stoichiometry of the cloned TauT was 2:1:1, and the transporter was specific for taurine and other β-amino acids with a high affinity for taurine [19].

2.2. Mechanisms Regulating the Activity of the Taurine Transporter

Taurine transporter activity was suppressed by exposure to taurine in JAR choriocarcinoma cells by both transcriptional and posttranscriptional mechanisms [20]. Another group also reported that TauT activity was downregulated by taurine itself in human intestinal Caco-2 cells, and this adaptive regulation was also induced by hypotaurine and β-alanine [21].

As for transcriptional regulatory mechanisms, TauT gene expression was elevated under high nitric oxide (NO) level conditions in retinal pigment epithelial cells [22]. Wilms tumor suppressor gene (WT1) enhances the transcription of TauT in human embryonic kidney 293 cells in a concentration-dependent manner [23]. Taurine transporter activity and the expression of TauT mRNA were upregulated by hypertonicity in mouse 3T3-L1 adipocytes [24]. Exposure to various prooxidants, including H_2O_2, induced the promoter activity of TauT in human retinal pigment epithelial cells [25]. Myoblast determination protein 1 (often abbreviated as MyoD) overexpression induced TauT promoter activity in mouse C3H10T1/2 fibroblasts [26]. Glucose reduced TauT mRNA and protein expression in a concentration-dependent manner in human Schwann cells [27].

The acute regulation of TauT activity involves a shift in pH and membrane potential (uptake is reduced following acidification of the extracellular medium and depolarization of the plasma membrane) and phosphorylation/dephosphorylation of TauT [28]. The cloned retinal TauT expressed in Xenopus oocytes is inhibited by the activation of cAMP-dependent protein kinase (PKA) and PKC [29]. Taurine uptake was downregulated by applying a PKC activator in the rat conditionally immortalized STB cell line (TR-TBT cells) [30]. Taurine uptake was inhibited, and efflux was accelerated under calcium-free conditions in TR-TBT cells [30]. Conditionally immortalized rat brain capillary endothelial cells (TR-BBB13), an in vitro model of the blood–brain barrier (BBB), showed an enhanced uptake of taurine by exposure to both tumor necrosis factor α (TNF-α) and a hypertonic condition [31]. Mochizuki et al. also showed that TNF-α markedly enhanced TauT activity in human intestinal Caco-2 cells. They proposed that the enhanced transport of taurine into cells underlies the cellular response against intestinal inflammation [32]. They further showed that nuclear factor κB (NF-κB) was involved in the signaling mechanism for the upregulation of taurine uptake and TauT mRNA expression induced by TNF-α [33]. Another group reported that TNF-α, lipopolysaccharide (LPS), and diethyl maleate (DEM) significantly increase taurine uptake, but H_2O_2 and an NO donor decreased taurine uptake in the cells [30]. Osmolarity also matters in the regulation of TauT activity. Upregulation of TauT mRNA transcription and TauT activity by hypertonic conditions in Caco-2 cells depends on calcium/calmodulin-dependent protein kinase II (Ca^{2+}/CaM kinase II) [34]. Cell swelling induced by hypoosmotic conditions results in reactive oxygen species production, which accounts for reduced taurine uptake under low-sodium/hypo-osmotic conditions by the direct modulation of TauT in NIH3T3 mouse fibroblasts [35]. Elevated levels of cortisol and IGF-1 also upregulate taurine transport in L6, a rat skeletal muscle cell line [36].

In summary, taurine itself downregulates TauT activity by transcriptional and post-transcriptional mechanisms. NO, hypertonicity, prooxidants, and glucose can upregulate TauT mRNA transcription. The factors that acutely enhance taurine transport by TauT include hypertonic condition, TNF-α, cortisol, LPS, and DEM. The factors that acutely inhibit taurine's transport by TauT include high glucose levels, hypoosmotic condition, and H_2O_2. In addition, NO enhances TauT activity in retinal pigment epithelial cells, while NO downregulates taurine uptake in TR-TBT cells.

2.3. Transfer of Taurine to the Fetus via the Placenta

2.3.1. Machinery for Taurine Transport in the Placenta

Sturman et al. have shown that [^{35}S]-labeled taurine, injected intraperitoneally into pregnant rats, can be delivered to both the brain and liver of the fetus, suggesting that maternal taurine can be transmitted to the fetus via the placenta [37]. Taurine is ranked as the most abundant free amino acid in the human placenta [38]. The human placenta transports taurine from the mother to the fetus by an active process because the taurine concentration in fetal blood is higher than in maternal blood [14]. Hibbard et al. demonstrated that the perfused human placenta can achieve and maintain that the ratio of fetal to maternal taurine concentration is, on average, 1.38:1, suggesting that taurine is transported in the placenta by an active transport mechanism [39]. Furthermore, the placental tissue concentration of taurine is 100 to 150-fold higher than that in fetal and maternal circulations [14]. Syncytiotrophoblast (STB) cells in the placenta possess an active transport system for taurine [40]. Kulanthaivel et al. found that the JAR human placental choriocarcinoma cell line can transport taurine, concentrating over 1000-fold inside the cell from the medium [14]. STB represents the primary barrier for transferring nutrients from the mother to the fetus in the human placenta. Maternal blood accumulates in the intervillous space and bathes the microvillous membrane (MVM). The basal plasma membrane (BM) of the STB is oriented toward fetal circulation. Transporters transferring amino acids, glucose, and fatty acids are expressed in both plasma membranes of the STB [41]. Taurine transport to the fetus also includes its uptake from maternal blood by transfer across the MVM of STB and, subsequently, they are transported to the fetus across BM [42]. The tissue concentrations of taurine in the human placenta are 100 to 200-fold higher than that in maternal blood, indicating the presence of an efficient active transport of taurine in the MVM [42]. Norberg et al. demonstrated that the activity of Na^+-dependent taurine transport, which is based on TauT, in BM was only 6% of that in MVM. In contrast, Na^+-independent transport activities also exist both in MVM and BM, although the details for Na^+-independent taurine transport remain to be elucidated [42]. The highly polarized distribution of Na^+/taurine cotransporters in the MVM in conjunction with similar Na^+-independent transport rates for taurine in MVM and BM provides the basis for net taurine flux between the mother and the fetus [42].

As for species' differences, the expression levels of TauT proteins were slightly lower in plasma membrane vesicles isolated from the mouse placenta than in human MVM, suggesting that the relative significance of placental taurine transport for fetal development is lower in mice than in humans [43]. However, the importance of the system β-amino acid transporter, encoded by the TauT gene, in promoting normal growth is exemplified by the observation that mice homozygous for the deletion of the TauT gene (taut−/−) are significantly smaller than their wild-type siblings and do not exhibit catch-up growth [43,44].

Amniotic fluid (AF) contains growth factors and principal nutrients that facilitate fetal growth and provide mechanical cushioning [45]. AF contains taurine, which is found in greater quantities in AF than in maternal serum. At the same time, most other amino acids have lower concentrations in AF than in maternal and fetal blood, indicating the activity of an unidentified mechanism by which taurine is enriched in AF [45].

In summary, the placental tissues concentrate taurine efficiently and transfer taurine to fetal circulation by Na^+-dependent taurine transport, which is based on TauT activity. Although the detailed mechanisms remain to be elucidated, taurine is also enriched in AF to a great extent. These observations suggest that taurine is essential in mother–fetus relationships, and taurine transport has to be adequately regulated during pregnancy.

2.3.2. Environmental Factors Affecting Placental Taurine Transport

In intrauterine growth restriction (IUGR), MVM Na^+-dependent taurine transport was found to be reduced by up to 34%, whereas Na^+-independent uptake was unaltered, suggesting that the impairment in placental taurine transport can be a causal factor of IUGR [42]. Roos et al. found that MVM TauT expression was unaltered in IUGR, whereas NO release downregulated placental TauT activity [46]. Ditchfield et al. showed that the placental TauT activity was significantly lower in obese women (body mass index (BMI) > 30) than women of ideal weight [47]. TauT activity in STB, measured in fragments of placental tissue, was negatively correlated with maternal BMI [48]. STB TauT activity was significantly lower in preeclampsia (PE) than in normal pregnancy [48]. In baboons, maternal nutrient restriction reduced fetal weight by 13%, decreased taurine concentration in fetal serum, and decreased TauT protein expression in MVM [49].

Preeclampsia is associated with NO signaling [50]. Therefore, the observation that PE is associated with reduced TauT activity is consistent with the observation that NO decreased taurine uptake in immortalized STB cell line TR-TBT [30]. High glucose levels inhibit TauT activity [27]. It can be inferred that PE and obesity are associated with reduced placental TauT activity because high blood glucose levels are well associated with both physiological conditions [51,52]. It is well known that, in obesity, adipose tissues release inflammatory mediators such as TNF-α and interleukin-6, predisposing the system to a pro-inflammatory state and to oxidative stress [47,53]. Severe social stress and chronic hypertension can act in combination to increase the risk of PE [54]. However, as described, TNF-α was observed to enhance taurine transport by TauT in the study using TR-TBT cells [30], and cortisol enhances TauT activity in a muscle cell line [36]. These in vitro observations are not apparently consistent with reduced TauT activity in the placenta in both obesity and PE. The detailed mechanisms for the downregulation of placental TauT activity in obesity and PE need further investigation.

Maternal environmental factors including obesity, PE, and nutrient restriction can suppress placental taurine transport. Environmental factors affecting placental taurine transfer can finally influence fetal development. These environmental factors are well known for their associations with undesirable developmental outcomes. For example, maternal obesity can be a risk factor for various mental disorders by affecting the development of the brains and neural circuits of the offspring [55]. Obstetric complications, including PE and nutrient deficiencies, also increase the risks for psychiatric disorders by inducing abnormal brain development during prenatal and postnatal periods [56]. Taurine might be a critical factor in these causal relationships.

2.4. Transfer of Taurine to Offspring during Postnatal Care

2.4.1. Transfer of Taurine to Offspring via Breast Milk

Exclusive breastfeeding, the practice of only providing breast milk for the first 6 months of an infant's life (no other food or water) provides essential, irreplaceable nutrition for a child's growth and development [57]. Taurine is a significant component of free amino acids in the milk of many species and is second in concentration to glutamate [10]. Taurine concentrations in chimpanzees, rhesus monkeys, sheep, and rats were found to be the highest during the first days of lactation, and it decreased to a particular constant concentration after the first week [10]. Sturman et al., by injecting [^{35}S]taurine intraperitoneally into lactating dams after parturition, demonstrated that taurine was transferred to pups via milk and accumulated in brains to a greater extent than in livers [58].

Milk production in the mammary gland mainly depends on milk synthesis and the proliferation abilities of mammary epithelial cells (MECs) [59,60]. Milk synthesis is the combined result of several intracellular processes within MECs [61]. Proteins synthesized in the endoplasmic reticulum of MECs are packaged into vesicles within the Golgi apparatus and then released by exocytosis. Some vesicles containing other proteins, such as IgA, are transported across the apical membrane. Some monosaccharides, sodium, potassium, chloride, and water can directly pass through the apical membrane [61]. Under the influence of progesterone and prolactin, MECs differentiate into the lobuloalveolar complex: a single layer of polarized MECs surrounding a lumen connected to the central duct system [61]. With the fall of progesterone at the end of pregnancy, when tight junctions form between MECs, milk comes to be contained within the lumen of the lobuloalveolar complex, and it becomes available for secretion [61].

Taurine transporters are expressed in mammary glands [62]. Aleman et al. showed that TauT mRNA was abundant during pregnancy in rat mammary glands. However, the expression levels decreased after the onset of lactation and stabilized around the levels observed in virgin rats [62]. Another study in mice reported lower transcription levels of TauT during lactation compared to the levels during pregnancy [63]. As shown in Figure 1, taurine was synthesized from cysteine via oxidation of cysteine to cysteine sulfinic acid by cysteine dioxygenase (CDO), followed by the decarboxylation of cysteine sulfinic acid to hypotaurine, a precursor of taurine, which is catalyzed by CSAD [64]. The mRNA transcription level for CSAD, a rate-limiting enzyme for taurine synthesis, was higher during early lactation (day 1 and 6 of lactation) than in the later lactational stage (day 14) in rat mammary glands [65]. CSAD mRNA was observed to be expressed in MECs in rat mammary glands [65]. The expression of CDO proteins was observed, preferentially, in the ductal cells of pregnant rats but not in other MECs or the ductal cells of nonpregnant rats [64]. In rats, milk taurine concentrations were the highest in early lactation, right after birth, and declined rapidly after the onset of lactation [64,66]. The expression levels of CDO proteins in the mammary tissue increased, and CSAD protein expression levels declined only slightly throughout lactation [64]. These results suggest that, in addition to taurine transported by MECs from maternal blood, a significant amount of taurine contained in mother's milk may be synthesized de novo in MECs [62].

2.4.2. Taurine in Infant Formula

As described, in the liver and brains of human fetuses and newborn infants, the levels of taurine biosynthesis are scant due to the limited activities of γ-cystathionase and CSAD [8,67,68]. In addition, human milk contained a large amount of taurine (450 to 500 mg/L), which resulted in the widely accepted notion that taurine functions as semi-essential amino acids [69]. These observations and widely accepted notions have supported the decision by the US Food and Drug Administration to permit adding taurine to purified infant formulas [69]. Analyses on free amino acids included in the commercially available formulas in the 2000s in Europe revealed that the reconstituted infant formulae contained a mean of 4.0 mg taurine/100 mL, and the reconstituted follow-up milk contained 1.8 mg taurine/100 mL [70]. The mean content of taurine found in infant formulae was similar to that found in human milk [70]. A recent systematic review found a complete lack of scientific evidence with regards to the developmental benefits of adding taurine to infant formula [71]. Therefore, further clinical studies are necessary for the evaluation of the beneficial effects of taurine in infant formulas.

2.4.3. Environmental Factors Affecting Taurine Transfer during Postnatal Period via Breast Milk

Intraperitoneal injection of taurine into lactating mouse dams did not affect taurine concentration in breast milk, but the injection of β-alanine significantly decreased taurine concentrations [72]. β-alanine administration caused a significant reduction in taurine concentration in the brains of offspring, and the taurine concentration in the brain was negatively correlated with the total distance traveled in the open field test at postnatal day 15, suggesting that a decreased concentration of taurine in the mother's milk can alter offspring behavior [72]. Interestingly, restraint stress in lactating mice caused an increase in the concentration of taurine and cystathionine in milk. However, restraint stress did not alter their concentration in the maternal plasma, liver, and mammary glands [73]. The ratio of taurine concentration in breast milk to its concentration in maternal plasma was significantly increased in the restraint stress group, suggesting that maternal stress promoted taurine transportation from maternal blood into milk [73]. Although the detailed molecular mechanisms underlying this phenomenon remain to be elucidated, the enhanced transport of taurine into milk under maternal stress conditions may be associated with the upregulation of the activity of TauT by the stress hormone, cortisol, at the cellular level as described above [36].

Summarizing the environmental factors relevant to taurine transport to offspring via mother's milk, excessive β-alanine administration can be a factor that decreases taurine concentration in breast milk, while maternal stress may enhance taurine transport to milk.

3. Biological Functions of Taurine during Development

3.1. Developmental Outcomes of Taurine-Depletion

The crucial observations related to the biological outcomes induced by taurine depletion were obtained in cats and monkeys. Cats fed a casein diet exhibited retinal degeneration [12]. Plasma amino acid profiles demonstrated an essential absence of plasma taurine in taurine-depleted cats. In contrast, concentrations of other amino acids, including methionine and cystine (taurine precursors), were comparable to the control values [12]. Adult female cats fed a defined taurine-free diet exhibited a low fertility rate and suffered retinal degeneration [74]. Except for taurine content, the maternal milk of the taurine-depleted mothers remained unchanged [74]. The concentration of taurine in the milk from taurine-deprived feline mothers was reduced to 5.9% of that from taurine-supplemented dams [5,13]. The surviving offspring from the taurine-depleted mothers exhibited a variety of neurological abnormalities and reduced concentrations of taurine in body tissues and fluids [74]. Neuringer and Sturman fed rhesus monkeys a taurine-free, soy protein-based infant formula from birth to 3 months of age, which resulted in impaired visual acuity [75]. Hayes and colleagues raised infant monkeys from birth with soybean infant milk formula lacking taurine and found significant growth depression, suggesting that dietary taurine is essential for infant development [76].

TauT knockout (KO) mice exhibited reduced fertility and retinal degeneration [77]. In several other studies, TauT KO mice exhibited a deficiency in myocardial and skeletal muscle taurine content compared to their wild-type littermates [44,78,79]. The TauT KO heart was characterized by a reduction in ventricular wall thickness and cardiac atrophy accompanied by smaller cardiomyocytes [78]. The skeletal muscles of TauT KO mice also exhibited decreased cell volume, structural defects, and a reduction in exercise endurance [78–81]. The expression of molecules related to osmotic stress was upregulated in both the heart and skeletal muscle of the TauT KO mice [78]. Tissue taurine depletion in TauT KO mice shortened the lifespan and accelerated the histological and functional defects of skeletal muscles, probably caused by cellular senescence [81]. Unfolded protein response was more often observed in TauT KO muscles than in control muscles, suggesting that taurine depletion causes an accumulation of protein misfolding, which accelerates cellular aging [81].

3.2. Possible Biological Mechanisms Underlying the Function of Taurine during Development

3.2.1. Taurine as Agonist for Receptors

Taurine is one of the most plentiful free amino acids in the developing central nervous system [82,83]. Extracellular taurine stimulates neurons and neuronal progenitor cells mainly by way of $GABA_A$ with affinities for specific receptor subtypes [83–85]. Taurine functions also as an agonist for $GABA_B$ receptors [84,86,87]. Taurine also functions as a partial agonist of glycine receptors [84]. Furthermore, neurons undergo specific maturation steps during brain development, including neurogenesis, neuronal migration, and neuronal anatomical and functional maturation, where taurine can function as extrinsic instructive signals for the neural progenitors (NPs) and the newly generated neurons in the developing central nervous system (CNS) [84].

During neurogenesis, NPs in a specific region of the CNS produce specific types of neurons in a defined order with precise timing throughout the course of CNS development [88,89]. The temporal and spatial specifications of NPs are essential for CNS histogenesis [89]. Temporal changes in NP differentiation are driven by a combination of intrinsic cellular properties and extracellular signals from the environment of the developing brain [90]. Neural progenitors express $GABA_A$ receptors in the developing cortex [91,92]. In the mouse developing cerebral cortex, Tochitani et al. demonstrated that taurine functions as an agonist for $GABA_A$ receptors to instruct temporal specifications of the NPs producing excitatory glutamatergic neurons [93]. Neural progenitors respond to taurine before embryonic day 13 (E13) through $GABA_A$ receptors, while NPs respond to GABA only after E13 [93]. Furthermore, the endogenous sources for non-synaptic GABA, such as the meninges and choroid plexus, are not fully developed before E13 in the developing cortex [94]. The taurine concentration was almost 500-times higher than GABA concentrations in the telencephalon at E13 [93]. $GABA_A$ antagonist administration to pregnant dams on E10–12, at which taurine functions as a principal endogenous agonist for $GABA_A$ receptors, resulted in the development of autistic behavior in offsprings [93]. These results not only demonstrate that taurine plays a vital role as an agonist for $GABA_A$

receptors but also that disruptions of taurine-GABA$_A$ receptor interaction can result in neurodevelopmental disorders.

Aberrant layer formation in the cerebrum and visual cortex was observed in the surviving kitten from taurine-depleted dams, suggesting that taurine plays a role in neuronal migration during brain development [13,95]. Behar et al. have shown that taurine and GABA$_B$ receptors mediate motility signals for the radial migration of newly generated neurons from the ventricular zone to the cortical plate in the developing rat cortex [96]. Recently, Furukawa et al. found that radial migration was not affected in a homozygous Glutamic acid decarboxylase 67 (GAD-67) deficient mouse, in which GABA content was reduced to 12.7% of the wild-type level [84,97]. However, the inhibition of GABA$_A$ receptors accelerated radial migration in this GAD-67-deficient mouse to a similar extent as in the wild-type animals, indicating that GABA is not required as an endogenous agonist for GABA$_A$ receptors regulating migration [84,97]. Furthermore, they observed that the radial migration of newly generated neurons was accelerated by maternal administration of D-cysteine sulfinic acid, a competitive inhibitor of taurine production, in both wild-type and homozygous GAD-67 deficient mice, suggesting that taurine is indeed a major endogenous modulator of radial migration [84,97]. Glycine receptors also participate in regulating the radial migration of newly generated neurons. Nimmervoll et al. showed that the application of strychnine, an antagonist of glycine receptors, retarded radial migration [98]. The application of strychnine without glycine did not cause any effects, suggesting that endogenous glycine does not induce the phenomenon. However, it remains unelucidated whether taurine can function as an endogenous agonist for glycine receptors to regulate the migration of newly generated neurons [98].

In the developing cortex, taurine accumulates in the cells in specific layers called the marginal zone (MZ) and subplate (Figure 2) [93,97]. A few cells in the cortical plate were also observed to contain high concentrations of taurine [93]. Extracellular taurine is released from cells through volume-regulated anion channels [97]. The MZ is the most superficial layer composed of the early generated neurons in the developing cortex (Figure 2). Cajal–Retzius cells in the MZ express glycine receptors and the activation of glycine receptors in the MZ results in membrane depolarization [99,100]. Qian et al. showed that a single electrical stimulation in MZ leads to the radial propagation of membrane depolarization in MZ, which can be suppressed by either GABA$_A$ or glycine receptor antagonists [101]. Qian et al. further showed that electrical stimulations induced the release of taurine and GABA, but not glycine, from the cells in MZ, suggesting that the activity-dependent release of taurine by cells in MZ mediates excitatory neurotransmission via GABA$_A$ and glycine receptors [101]. Flint et al. showed that the immature postmitotic neurons in the intermediate zone and the cortical plate in the developing cortex express functional glycine receptors, which are excitatory, while neural progenitors in the ventricular zone do not express functional glycine receptors [102]. Flint et al. also found that the extracellular accumulation of taurine induced by applications of guanidinoethylsulfinate, an inhibitor of sodium-taurine cotransport, produced large inward currents that were reversibly blocked by strychnine, an inhibitor of glycine receptors, suggesting that taurine functions as an endogenous agonist for glycine receptors in the immature postmitotic neurons in the developing cortex [102].

As previously mentioned, AF contains taurine in greater concentrations than in maternal serum. The AF trapped in the neural tube serves as the initial cerebrospinal fluid (CSF) during neural tube closure [103,104]. The cells facing the lateral ventricle were also rich in taurine in addition to the cells in the mantle zone and subplate, especially in the early phase of cortical development, suggesting that the CSF contained in the ventricles may be a source of extracellular taurine in the developing cortex [93]. Indeed, the injection of taurine into the AF rescued the cortical phenotypes of TauT KO embryos [93].

In summary, extracellular taurine functions as an agonist for GABA$_A$ receptors, GABA$_B$ receptors, and glycine receptors. Extracellular taurine is released from the cells via volume-sensitive channels or is present in CSF, which originates from AF in the initial phase of brain development, contained in the ventricles of the developing brains.

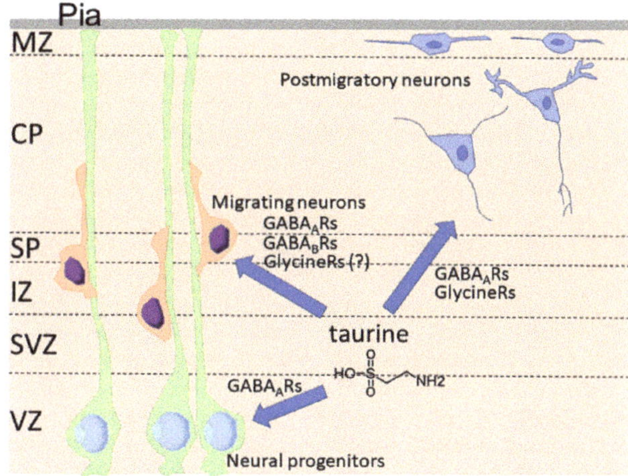

Figure 2. Functions of taurine as endogenous agonists for $GABA_A$ receptors ($GABA_A$Rs), $GABA_B$ receptors ($GABA_B$Rs), and glycine receptors (Glycine Rs) in the developing cortex. CP, cortical plate; IZ, intermediate zone; MZ, marginal zone; Pia, pia mater; SP, subplate; SVZ, subventricular zone; VZ, ventricular zone.

3.2.2. Other Mechanisms Related to Various Physiological Functions of Taurine during Development

Taurine possesses various cellular and physiological functions [9]. Taurine functions as a membrane stabilizer [105], is involved in cell volume regulation [9,106] and cellular calcium homeostasis [84,107], is incorporated into modified uridines in mitochondrial tRNAs [108–111], exhibits robust antioxidant effects [9,112–114], modulates inflammation and apoptosis [9,115–117], and promotes tissue repair in combination with branched-chain amino acids [118]. These basic but vital functions of taurine at the cellular level must support normal systemic development. Some of the various phenotypes observed in taurine depletion experiments could be related to these principal cellular functions, although further studies are needed from such a point of view.

3.2.3. Function of Taurine as a Factor to Shape the Gut Microbiota

In the last division of this section, I will discuss a lesser acknowledged although potentially critical role of taurine. The colonization of gut microbiota in the offspring relies on the maternal resident flora as a primary source of the microbiome during perinatal periods [119]. Microbes colonize human and other mammalian bodies during the first moments of life and coexist with the host throughout later life [120]. Many lines of studies have demonstrated that gut microbiota play significant roles in the regulation of the physiology of the hosts [121]. The gut microorganisms that reside in the host during development can affect the overall development of the host [119,122,123].

Bile acids (BAs) are the multifunctional products of cholesterol metabolism that occur in a wide range of vertebrates and aid in the absorption of fats and fat-soluble vitamins from the diet [119,124,125]. BAs are synthesized in the hepatocytes of the liver and secreted into bile. Most BAs are actively absorbed by the ileum and enter the portal vein [124]. After their synthesis in the hepatocytes, primary BAs are conjugated by N-acylamidation with taurine or a taurine-derivative or, less commonly, with glycine in rodents [124]. After conjugated BAs are secreted into the intestinal tract, they are modified by the gut microbiota to produce secondary BAs, which lack taurine or glycine residues [119,126]. The synthesis of BAs is subject to negative feedback control, which is modulated by the nuclear receptor farnesoid X receptor (FXR) in the ileum and liver [127]. By comparing BA profiles in germ-free (GF) mice and in conventionally raised (CONV-R) mice, Sayin demonstrated that the gut microbiota not only regulated the secondary bile acid metabolism but also modified bile acid synthesis in the liver by modulating FXR signaling [127]. The gut microbiota modulate BA synthesis by regulating the expression of the enzymes in hepatocytes involved in BA synthesis [119,128]. In other words, BAs synthesized by the host influence intestinal bacterial compositions, and intestinal bacteria conversely modify the circulating BA composition in the host [119,128]. Notably, Sayin et al. also found that taurine levels in the liver decreased in CONV-R mice than in GF mice, while the expression levels of CSAD and TauT increased in the livers of CONV-R mice compared to those of GF mice [127]. Although the detailed underlying mechanisms for enhancing hepatic taurine biosynthesis and transport in CONV-R mice remain unknown, it can be at least inferred that the homeostatic mechanism for taurine may be involved in gut microbiota–BA interactions.

Miyazaki et al. analyzed BA profiles in taurine-depleted cats and demonstrated that the total BA concentration in bile was higher in the control than that in the taurine-depleted group [129]. However, the total BA concentration in serum was lower in the control than that in taurine-depleted cats. BA profiles in bile also differed between control and taurine-depleted cats [129]. The authors explained the mechanism as follows. Impaired BA metabolism in the liver is induced by decreased mitochondrial cholesterol 27-hydroxylase expression and mitochondrial activity [129]; taurine depletion decreases taurine-modified mt-tRNAs, resulting in reduced expressions of CYP27A1, a key enzyme for primary BA synthesis. BA synthesis, particularly in chenodeoxycholic acid synthesis pathway, decreases [129]; in addition, BA excretion into bile is decreased, and, conversely, excretion into the peripheral circulation is increased [129]. These results illustrate the function of taurine in BA homeostasis. Recently, Stacy et al. revealed a taurine-based mechanism in which BAs influence gut microbiota, resulting in enhanced resistance to pathogens [130,131]. The authors found that the gut microbiota from previously infected hosts displays enhanced the resistance to infections. This long-term functional remodeling of the host's immunity is associated with altered BA metabolism resulting in the expansion of bacterial taxa that utilize taurine [130]. Interestingly, supplying exogenous taurine alone is sufficient for inducing alterations in microbiota function and enhancing resistance to the pathogens [130]. Further analyses showed that taurine potentiates microbial production of sulfide, an inhibitor of cellular respiration, which plays a protective role against the invasion of the host by numerous pathogens [130]. This suggests that taurine sustains and trains the microbiota, which promotes host resistance to subsequent infection [130].

In summary, taurine can directly influence the microbiome by its potential as a sulfide supplier or by its ability to conjugate with primary BAs. Therefore, taurine could indirectly affect the development of the host by shaping resident microbiota. Such functions should be addressed in future studies.

3.3. Possible Influence of Limited Levels of Taurine during Development on Disease Risk in Adults

Low birth weight relates closely to the increased incidence of coronary heart disease and related disorders, such as stroke, hypertension, and adult-onset diabetes [132]. A new 'developmental' hypothesis for the etiology considers specifically how development in early life affects the development of chronic diseases later in life [132,133]. The theory was proposed by Barker and now developed to the concepts of *developmental programming* or *the developmental origins of the health and disease hypothesis (DOHaD)* [134,135]. According to the hypothesis, exposure in early life to environmental factors, such as maternal nutrition, infant feeding methods, maternal stress, and infection, influences the long-term risk of various diseases [135,136]. Therefore, IUGR exerts its long-term effects on disease susceptibility later in life [42,137]. As discussed earlier in this review, IUGR is associated with reduced TauT activity, and, importantly, low plasma concentrations of taurine are often found in IUGR fetuses [42,46]. Reduced placental TauT activity is associated with maternal obesity and PE [47,48]. Similarly, IUGR is often associated with maternal obesity and PE [138,139]. Furthermore, maternal nutrient restriction reduced placental TauT expression and the concentration of amino acids, including taurine, which restricted the growth of the fetus [49]. These results suggest that reduced taurine levels, induced by, among other factors, maternal obesity, PE, and malnutrition during fetal and infant development can result in limited body growth and could increase the risks of various chronic diseases in later life. Future studies are strongly needed from such a point of view.

4. Conclusions

Taurine transport from mother to offspring underlies the functioning of various physiological factors that determine healthy development. Taurine transfer via the placenta can be inhibited by various environmental factors, including maternal obesity, PE, and malnutrition. In breast milk, maternal stress increases taurine concentration, while excessive maternal β-alanine ingestion results in a decrease in the concentration of taurine (Figure 3). Taurine depletion has diverse adverse effects on offspring development; further studies are needed concerning the regulatory mechanisms of taurine transport, the environmental factors that influence the activity of taurine transport, and the immediate and long-term health outcomes associated with limited taurine availability during prenatal and perinatal periods.

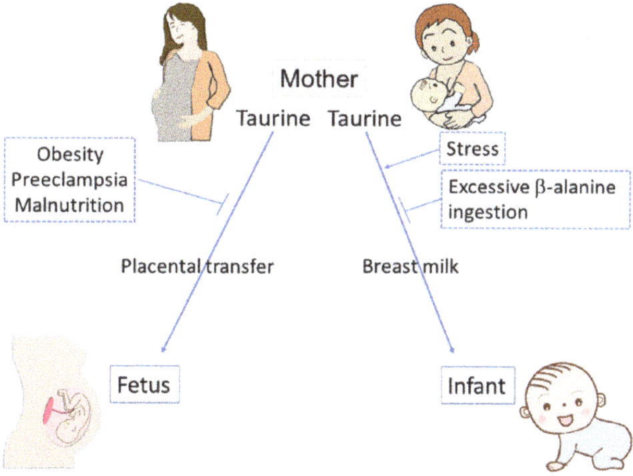

Figure 3. Taurine transfers from mother to offspring via either the placenta or breast milk and several environmental factors affect the transfer of taurine. Maternal obesity, preeclampsia, and malnutrition can inhibit the placental transfer of taurine. Excessive β-alanine ingestion can inhibit taurine transfer via breast milk, while maternal stress can enhance it.

Funding: This work was supported by KAKENHI (Grant Numbers; 26461629, 17H04654, 17K19225, 20K20320, and 20K08242) from the Japan Society for the Promotion of Science (JSPS); a research grant from OKASAN-KATO Foundation, Mie, Japan; a research grant from the Morinaga Foundation for Health & Nutrition, Tokyo, Japan; and a research grant from the Mishima Kaiun Memorial Foundation, Tokyo, Japan.

Acknowledgments: The author wishes to thank Atsuo Fukuda (Hamamatsu University School of Medicine) and the various researchers that provided valuable input and feedback and sacrificed their time for intellectual discourse. The author also wishes to acknowledge a large number of relevant and essential studies that could not be mentioned or discussed in this review.

Conflicts of Interest: The author declares no conflict of interest.

References

1. Broad, K.D.; Curley, J.P.; Keverne, E.B. Mother-infant bonding and the evolution of mammalian social relationships. *Philos. Trans. R. Soc. Lond. B Biol. Sci.* **2006**, *361*, 2199–2214. [CrossRef] [PubMed]
2. Mogi, K.; Nagasawa, M.; Kikusui, T. Developmental consequences and biological significance of mother-infant bonding. *Prog. Neuro-Psychopharmacol. Biol. Psychiatry* **2011**, *35*, 1232–1241. [CrossRef] [PubMed]
3. Daglar, G.; Nur, N. Level of mother-baby bonding and influencing factors during pregnancy and postpartum period. *Psychiatr Danub.* **2018**, *30*, 433–440. [CrossRef]
4. Wu, J.Y.; Prentice, H. Role of taurine in the central nervous system. *J. Biomed. Sci.* **2010**, *17* (Suppl. S1), S1. [CrossRef] [PubMed]
5. Tochitani, S. Functions of Maternally-Derived Taurine in Fetal and Neonatal Brain Development. *Adv. Exp. Med. Biol.* **2017**, *975 Pt 1*, 17–25. [CrossRef]
6. Linne, M.L.; Jalonen, T.O.; Saransaari, P.; Oja, S.S. Taurine-induced single-channel currents in cultured rat cerebellar granule cells. *Adv. Exp. Med. Biol.* **1996**, *403*, 455–462. [PubMed]
7. Ye, G.; Tse, A.C.; Yung, W. Taurine inhibits rat substantia nigra pars reticulata neurons by activation of GABA- and glycine-linked chloride conductance. *Brain Res.* **1997**, *749*, 175–179. [CrossRef]
8. Sturman, J.A.; Hayes, K.C. The Biology of Taurine in Nutrition and Development. *Adv. Nutr. Res.* **1980**, 231–299. [CrossRef]
9. Lambert, I.H.; Kristensen, D.M.; Holm, J.B.; Mortensen, O.H. Physiological role of taurine–from organism to organelle. *Acta Physiol.* **2015**, *213*, 191–212. [CrossRef] [PubMed]
10. Sturman, J.A.; Rassin, D.K.; Gaull, G.E. Taurine in development. *Life Sci.* **1977**, *21*, 1–22. [CrossRef]
11. Sturman, J.A. Origin of taurine in developing rat brain. *Brain Res.* **1981**, *254*, 111–128. [CrossRef]
12. Hayes, K.C.; Carey, R.E.; Schmidt, S.Y. Retinal degeneration associated with taurine deficiency in the cat. *Science* **1975**, *188*, 949–951. [CrossRef] [PubMed]
13. Sturman, J.A.; Moretz, R.C.; French, J.H.; Wisniewski, H.M. Taurine deficiency in the developing cat: Persistence of the cerebellar external granule cell layer. *J. Neurosci. Res.* **1985**, *13*, 405–416. [CrossRef] [PubMed]
14. Kulanthaivel, P.; Cool, D.R.; Ramamoorthy, S.; Mahesh, V.B.; Leibach, F.H.; Ganapathy, V. Transport of taurine and its regulation by protein kinase C in the JAR human placental choriocarcinoma cell line. *Biochem. J.* **1991**, *277 Pt 1*, 53–58. [CrossRef] [PubMed]
15. Shennan, D.B.; McNeillie, S.A.; Curran, D.E. Stimulation of taurine efflux from human placental tissue by a hypoosmotic challenge. *Exp. Physiol.* **1993**, *78*, 843–846. [CrossRef]
16. Uchida, S.; Kwon, H.M.; Yamauchi, A.; Preston, A.S.; Marumo, F.; Handler, J.S. Molecular cloning of the cDNA for an MDCK cell Na(+)- and Cl(−)-dependent taurine transporter that is regulated by hypertonicity. *Proc. Natl. Acad. Sci. USA* **1992**, *89*, 8230–8234. [CrossRef]
17. Liu, Q.R.; Lopez-Corcuera, B.; Nelson, H.; Mandiyan, S.; Nelson, N. Cloning and expression of a cDNA encoding the transporter of taurine and beta-alanine in mouse brain. *Proc. Natl. Acad. Sci. USA* **1992**, *89*, 12145–12149. [CrossRef]
18. Smith, K.E.; Borden, L.A.; Wang, C.H.; Hartig, P.R.; Branchek, T.A.; Weinshank, R.L. Cloning and expression of a high affinity taurine transporter from rat brain. *Mol. Pharm.* **1992**, *42*, 563–569.
19. Ramamoorthy, S.; Leibach, F.H.; Mahesh, V.B.; Han, H.; Yang-Feng, T.; Blakely, R.D.; Ganapathy, V. Functional characterization and chromosomal localization of a cloned taurine transporter from human placenta. *Biochem. J.* **1994**, *300 Pt 3*, 893–900. [CrossRef]
20. Jayanthi, L.D.; Ramamoorthy, S.; Mahesh, V.B.; Leibach, F.H.; Ganapathy, V. Substrate-specific regulation of the taurine transporter in human placental choriocarcinoma cells (JAR). *Biochim. Biophys. Acta* **1995**, *1235*, 351–360. [CrossRef]
21. Satsu, H.; Watanabe, H.; Arai, S.; Shimizu, M. Characterization and regulation of taurine transport in Caco-2, human intestinal cells. *J. Biochem.* **1997**, *121*, 1082–1087. [CrossRef] [PubMed]
22. Bridges, C.C.; Ola, M.S.; Prasad, P.D.; El-Sherbeny, A.; Ganapathy, V.; Smith, S.B. Regulation of taurine transporter expression by NO in cultured human retinal pigment epithelial cells. *Am. J. Physiol. Cell Physiol.* **2001**, *281*, C1825–C1836. [CrossRef] [PubMed]
23. Han, X.; Chesney, R.W. Regulation of taurine transporter gene (TauT) by WT1. *FEBS Lett.* **2003**, *540*, 71–76. [CrossRef]
24. Takasaki, M.; Satsu, H.; Shimizu, M. Physiological significance of the taurine transporter and taurine biosynthetic enzymes in 3T3-L1 adipocytes. *Biofactors* **2004**, *21*, 419–421. [CrossRef] [PubMed]
25. Nakashima, E.; Pop-Busui, R.; Towns, R.; Thomas, T.P.; Hosaka, Y.; Nakamura, J.; Greene, D.A.; Killen, P.D.; Schroeder, J.; Larkin, D.D.; et al. Regulation of the human taurine transporter by oxidative stress in retinal pigment epithelial cells stably transformed to overexpress aldose reductase. *Antioxid. Redox Signal.* **2005**, *7*, 1530–1542. [CrossRef]
26. Uozumi, Y.; Ito, T.; Hoshino, Y.; Mohri, T.; Maeda, M.; Takahashi, K.; Fujio, Y.; Azuma, J. Myogenic differentiation induces taurine transporter in association with taurine-mediated cytoprotection in skeletal muscles. *Biochem. J.* **2006**, *394*, 699–706. [CrossRef]
27. Askwith, T.; Zeng, W.; Eggo, M.C.; Stevens, M.J. Oxidative stress and dysregulation of the taurine transporter in high-glucose-exposed human Schwann cells: Implications for pathogenesis of diabetic neuropathy. *Am. J. Physiol. Endocrinol. Metab.* **2009**, *297*, E620–E628. [CrossRef]
28. Lambert, I.H.; Hansen, D.B. Regulation of taurine transport systems by protein kinase CK2 in mammalian cells. *Cell. Physiol. Biochem.* **2011**, *28*, 1099–1110. [CrossRef]
29. Loo, D.D.F.; Hirsch, J.R.; Sarkar, H.K.; Wright, E.M. Regulation of the mouse retinal taurine transporter (TAUT) by protein kinases in *Xenopus* oocytes. *FEBS Lett.* **1996**, *392*, 250–254. [CrossRef]
30. Lee, N.Y.; Kang, Y.S. Regulation of taurine transport at the blood-placental barrier by calcium ion, PKC activator and oxidative stress conditions. *J. Biomed. Sci.* **2010**, *17* (Suppl. S1). [CrossRef]

31. Kang, Y.S.; Ohtsuki, S.; Takanaga, H.; Tomi, M.; Hosoya, K.; Terasaki, T. Regulation of taurine transport at the blood-brain barrier by tumor necrosis factor-alpha, taurine and hypertonicity. *J. Neurochem.* **2002**, *83*, 1188–1195. [CrossRef]
32. Mochizuki, T.; Satsu, H.; Nakano, T.; Shimizu, M. Regulation of the human taurine transporter by TNF-alpha and an anti-inflammatory function of taurine in human intestinal Caco-2 cells. *Biofactors* **2004**, *21*, 141–144. [CrossRef]
33. Mochizuki, T.; Satsu, H.; Shimizu, M. Signaling pathways involved in tumor necrosis factor alpha-induced upregulation of the taurine transporter in Caco-2 cells. *FEBS Lett.* **2005**, *579*, 3069–3074. [CrossRef] [PubMed]
34. Satsu, H.; Manabe, M.; Shimizu, M. Activation of Ca^{2+}/calmodulin-dependent protein kinase II is involved in hyperosmotic induction of the human taurine transporter. *FEBS Lett.* **2004**, *569*, 123–128. [CrossRef] [PubMed]
35. Hansen, D.B.; Friis, M.B.; Hoffmann, E.K.; Lambert, I.H. Downregulation of the taurine transporter TauT during hypo-osmotic stress in NIH3T3 mouse fibroblasts. *J. Membr. Biol.* **2012**, *245*, 77–87. [CrossRef]
36. Park, S.H.; Lee, H.; Park, T. Cortisol and IGF-1 synergistically up-regulate taurine transport by the rat skeletal muscle cell line, L6. *Biofactors* **2004**, *21*, 403–406. [CrossRef] [PubMed]
37. Sturman, J.A.; Rassin, D.K.; Gaull, G.E. Taurine in developing rat brain: Maternal-fetal transfer of [^{35}S] taurine and its fate in the neonate. *J. Neurochem.* **1977**, *28*, 31–39. [CrossRef]
38. Desforges, M.; Parsons, L.; Westwood, M.; Sibley, C.P.; Greenwood, S.L. Taurine transport in human placental trophoblast is important for regulation of cell differentiation and survival. *Cell Death Dis.* **2013**, *4*, e559. [CrossRef]
39. Hibbard, J.U.; Pridjian, G.; Whitington, P.F.; Moawad, A.H. Taurine transport in the in vitro perfused human placenta. *Pediatric Res.* **1990**, *27*, 80–84. [CrossRef]
40. Miyamoto, Y.; Balkovetz, D.F.; Leibach, F.H.; Mahesh, V.B.; Ganapathy, V. Na^{+} + Cl^{-} -gradient-driven, high-affinity, uphill transport of taurine in human placental brush-border membrane vesicles. *FEBS Lett.* **1988**, *231*, 263–267. [CrossRef]
41. Lager, S.; Powell, T.L. Regulation of nutrient transport across the placenta. *J. Pregnancy* **2012**, *2012*, 179827. [CrossRef] [PubMed]
42. Norberg, S.; Powell, T.L.; Jansson, T. Intrauterine growth restriction is associated with a reduced activity of placental taurine transporters. *Pediatric Res.* **1998**, *44*, 233–238. [CrossRef] [PubMed]
43. Kusinski, L.C.; Jones, C.J.; Baker, P.N.; Sibley, C.P.; Glazier, J.D. Isolation of plasma membrane vesicles from mouse placenta at term and measurement of system A and system beta amino acid transporter activity. *Placenta* **2010**, *31*, 53–59. [CrossRef]
44. Warskulat, U.; Heller-Stilb, B.; Oermann, E.; Zilles, K.; Haas, H.; Lang, F.; Haussinger, D. Phenotype of the taurine transporter knockout mouse. *Methods Enzymol.* **2007**, *428*, 439–458. [CrossRef] [PubMed]
45. Underwood, M.A.; Gilbert, W.M.; Sherman, M.P. Amniotic fluid: Not just fetal urine anymore. *J. Perinatol. Off. J. Calif. Perinat. Assoc.* **2005**, *25*, 341–348. [CrossRef] [PubMed]
46. Roos, S.; Powell, T.L.; Jansson, T. Human placental taurine transporter in uncomplicated and IUGR pregnancies: Cellular localization, protein expression, and regulation. *Am. J. Physiol. Regul. Integr. Comp. Physiol.* **2004**, *287*, R886–R893. [CrossRef] [PubMed]
47. Ditchfield, A.M.; Desforges, M.; Mills, T.A.; Glazier, J.D.; Wareing, M.; Mynett, K.; Sibley, C.P.; Greenwood, S.L. Maternal obesity is associated with a reduction in placental taurine transporter activity. *Int. J. Obes.* **2015**, *39*, 557–564. [CrossRef] [PubMed]
48. Desforges, M.; Ditchfield, A.; Hirst, C.R.; Pegorie, C.; Martyn-Smith, K.; Sibley, C.P.; Greenwood, S.L. Reduced placental taurine transporter (TauT) activity in pregnancies complicated by pre-eclampsia and maternal obesity. *Adv. Exp. Med. Biol.* **2013**, *776*, 81–91. [CrossRef] [PubMed]
49. Kavitha, J.V.; Rosario, F.J.; Nijland, M.J.; McDonald, T.J.; Wu, G.; Kanai, Y.; Powell, T.L.; Nathanielsz, P.W.; Jansson, T. Down-regulation of placental mTOR, insulin/IGF-I signaling, and nutrient transporters in response to maternal nutrient restriction in the baboon. *FASEB J.* **2014**, *28*, 1294–1305. [CrossRef] [PubMed]
50. Sutton, E.F.; Gemmel, M.; Powers, R.W. Nitric oxide signaling in pregnancy and preeclampsia. *Nitric Oxide* **2020**, *95*, 55–62. [CrossRef]
51. Weissgerber, T.L.; Mudd, L.M. Preeclampsia and diabetes. *Curr. Diab. Rep.* **2015**, *15*, 9. [CrossRef]
52. Al-Goblan, A.S.; Al-Alfi, M.A.; Khan, M.Z. Mechanism linking diabetes mellitus and obesity. *Diabetes Metab. Syndr. Obes.* **2014**, *7*, 587–591. [CrossRef] [PubMed]
53. Ellulu, M.S.; Patimah, I.; Khaza'ai, H.; Rahmat, A.; Abed, Y. Obesity and inflammation: The linking mechanism and the complications. *Arch. Med. Sci.* **2017**, *13*, 851–863. [CrossRef] [PubMed]
54. Yu, Y.; Zhang, S.; Wang, G.; Hong, X.; Mallow, E.B.; Walker, S.O.; Pearson, C.; Heffner, L.; Zuckerman, B.; Wang, X. The combined association of psychosocial stress and chronic hypertension with preeclampsia. *Am. J. Obstet. Gynecol.* **2013**, *209*, 438.e1–438.e12. [CrossRef]
55. Rivera, H.M.; Christiansen, K.J.; Sullivan, E.L. The role of maternal obesity in the risk of neuropsychiatric disorders. *Front. Neurosci.* **2015**, *9*, 194. [CrossRef] [PubMed]
56. Schmitt, A.; Malchow, B.; Hasan, A.; Falkai, P. The impact of environmental factors in severe psychiatric disorders. *Front. Neurosci.* **2014**, *8*, 19. [CrossRef] [PubMed]
57. WHO. Breastfeeding Policy Brief. 2014. Available online: https://www.who.int/publications/i/item/WHO-NMH-NHD-14.7 (accessed on 1 March 2022).
58. Sturman, J.A.; Rassin, D.K.; Gaull, G.E. Taurine in developing rat brain: Transfer of [35S] taurine to pups via the milk. *Pediatric Res.* **1977**, *11*, 28–33. [CrossRef]

59. Yu, M.; Wang, Y.; Wang, Z.; Liu, Y.; Yu, Y.; Gao, X. Taurine Promotes Milk Synthesis via the GPR87-PI3K-SETD1A Signaling in BMECs. *J. Agric. Food Chem.* **2019**, *67*, 1927–1936. [CrossRef] [PubMed]
60. Boutinaud, M.; Guinard-Flament, J.; HélèneJammes. The number and activity of mammary epithelial cells, determining factors for milk production. *Reprod. Nutr. Dev.* **2004**, *44*, 499–508. [CrossRef]
61. Jonas, W.; Woodside, B. Physiological mechanisms, behavioral and psychological factors influencing the transfer of milk from mothers to their young. *Horm. Behav.* **2016**, *77*, 167–181. [CrossRef] [PubMed]
62. Aleman, G.; Lopez, A.; Ordaz, G.; Torres, N.; Tovar, A.R. Changes in messenger RNA abundance of amino acid transporters in rat mammary gland during pregnancy, lactation, and weaning. *Metabolism* **2009**, *58*, 594–601. [CrossRef] [PubMed]
63. Rudolph, M.C.; McManaman, J.L.; Phang, T.; Russell, T.; Kominsky, D.J.; Serkova, N.J.; Stein, T.; Anderson, S.M.; Neville, M.C. Metabolic regulation in the lactating mammary gland: A lipid synthesizing machine. *Physiol. Genom.* **2007**, *28*, 323–336. [CrossRef] [PubMed]
64. Ueki, I.; Stipanuk, M.H. Enzymes of the taurine biosynthetic pathway are expressed in rat mammary gland. *J. Nutr.* **2007**, *137*, 1887–1894. [CrossRef] [PubMed]
65. Hu, J.M.; Ikemura, R.; Chang, K.T.; Suzuki, M.; Nishihara, M.; Takahashi, M. Expression of cysteine sulfinate decarboxylase mRNA in rat mammary gland. *J. Vet. Med. Sci.* **2000**, *62*, 829–834. [CrossRef] [PubMed]
66. Hu, J.M.; Rho, J.Y.; Suzuki, M.; Nishihara, M.; Takahashi, M. Effect of taurine in rat milk on the growth of offspring. *J. Vet. Med. Sci.* **2000**, *62*, 693–698. [CrossRef] [PubMed]
67. Gaull, G.; Sturman, J.A.; Raiha, N.C. Development of mammalian sulfur metabolism: Absence of cystathionase in human fetal tissues. *Pediatric Res.* **1972**, *6*, 538–547. [CrossRef] [PubMed]
68. Zlotkin, S.H.; Anderson, G.H. The development of cystathionase activity during the first year of life. *Pediatric Res.* **1982**, *16*, 65–68. [CrossRef] [PubMed]
69. Chesney, R.W.; Helms, R.A.; Christensen, M.; Budreau, A.M.; Han, X.; Sturman, J.A. The role of taurine in infant nutrition. *Adv. Exp. Med. Biol.* **1998**, *442*, 463–476. [CrossRef]
70. Ferreira, I.M. Quantification of non-protein nitrogen components of infant formulae and follow-up milks: Comparison with cows' and human milk. *Br. J. Nutr.* **2003**, *90*, 127–133. [CrossRef]
71. Almeida, C.C.; Mendonca Pereira, B.F.; Leandro, K.C.; Costa, M.P.; Spisso, B.F.; Conte-Junior, C.A. Bioactive Compounds in Infant Formula and Their Effects on Infant Nutrition and Health: A Systematic Literature Review. *Int. J. Food Sci.* **2021**, *2021*, 8850080. [CrossRef] [PubMed]
72. Nishigawa, T.; Nagamachi, S.; Chowdhury, V.S.; Yasuo, S.; Furuse, M. Taurine and beta-alanine intraperitoneal injection in lactating mice modifies the growth and behavior of offspring. *Biochem. Biophys. Res. Commun.* **2018**, *495*, 2024–2029. [CrossRef] [PubMed]
73. Nishigawa, T.; Nagamachi, S.; Ikeda, H.; Chowdhury, V.S.; Furuse, M. Restraint stress in lactating mice alters the levels of sulfur-containing amino acids in milk. *J. Vet. Med. Sci.* **2018**, *80*, 503–509. [CrossRef] [PubMed]
74. Sturman, J.A.; Gargano, A.D.; Messing, J.M.; Imaki, H. Feline maternal taurine deficiency: Effect on mother and offspring. *J. Nutr.* **1986**, *116*, 655–667. [CrossRef]
75. Neuringer, M.; Sturman, J. Visual acuity loss in rhesus monkey infants fed a taurine-free human infant formula. *J. Neurosci. Res.* **1987**, *18*, 597–601. [CrossRef]
76. Hayes, K.C.; Stephan, Z.F.; Sturman, J.A. Growth depression in taurine-depleted infant monkeys. *J. Nutr.* **1980**, *110*, 2058–2064. [CrossRef] [PubMed]
77. Heller-Stilb, B.; van Roeyen, C.; Rascher, K.; Hartwig, H.G.; Huth, A.; Seeliger, M.W.; Warskulat, U.; Haussinger, D. Disruption of the taurine transporter gene (taut) leads to retinal degeneration in mice. *FASEB J.* **2002**, *16*, 231–233. [CrossRef] [PubMed]
78. Ito, T.; Kimura, Y.; Uozumi, Y.; Takai, M.; Muraoka, S.; Matsuda, T.; Ueki, K.; Yoshiyama, M.; Ikawa, M.; Okabe, M.; et al. Taurine depletion caused by knocking out the taurine transporter gene leads to cardiomyopathy with cardiac atrophy. *J. Mol. Cell. Cardiol.* **2008**, *44*, 927–937. [CrossRef] [PubMed]
79. Warskulat, U.; Flogel, U.; Jacoby, C.; Hartwig, H.G.; Thewissen, M.; Merx, M.W.; Molojavyi, A.; Heller-Stilb, B.; Schrader, J.; Haussinger, D. Taurine transporter knockout depletes muscle taurine levels and results in severe skeletal muscle impairment but leaves cardiac function uncompromised. *FASEB J.* **2004**, *18*, 577–579. [CrossRef] [PubMed]
80. Ito, T.; Oishi, S.; Takai, M.; Kimura, Y.; Uozumi, Y.; Fujio, Y.; Schaffer, S.W.; Azuma, J. Cardiac and skeletal muscle abnormality in taurine transporter-knockout mice. *J. Biomed. Sci.* **2010**, *17* (Suppl. S1), S20. [CrossRef]
81. Ito, T.; Yoshikawa, N.; Inui, T.; Miyazaki, N.; Schaffer, S.W.; Azuma, J. Tissue depletion of taurine accelerates skeletal muscle senescence and leads to early death in mice. *PLoS ONE* **2014**, *9*, e107409. [CrossRef] [PubMed]
82. Benitez-Diaz, P.; Miranda-Contreras, L.; Mendoza-Briceno, R.V.; Pena-Contreras, Z.; Palacios-Pru, E. Prenatal and postnatal contents of amino acid neurotransmitters in mouse parietal cortex. *Dev. Neurosci.* **2003**, *25*, 366–374. [CrossRef] [PubMed]
83. Huxtable, R.J. Taurine in the central nervous system and the mammalian actions of taurine. *Prog. Neurobiol.* **1989**, *32*, 471–533. [CrossRef]
84. Kilb, W.; Fukuda, A. Taurine as an Essential Neuromodulator during Perinatal Cortical Development. *Front. Cell. Neurosci.* **2017**, *11*, 328. [CrossRef] [PubMed]
85. Kletke, O.; Gisselmann, G.; May, A.; Hatt, H.; Sergeeva, O.A. Partial agonism of taurine at gamma-containing native and recombinant GABAA receptors. *PLoS ONE* **2013**, *8*, e61733. [CrossRef] [PubMed]

86. Kontro, P.; Oja, S.S. Interactions of taurine with GABAB binding sites in mouse brain. *Neuropharmacology* **1990**, *29*, 243–247. [CrossRef]
87. Smith, S.S.; Li, J. GABAB receptor stimulation by baclofen and taurine enhances excitatory amino acid induced phosphatidylinositol turnover in neonatal rat cerebellum. *Neurosci. Lett.* **1991**, *132*, 59–64. [CrossRef]
88. Gotz, M.; Huttner, W.B. The cell biology of neurogenesis. *Nat. Reviews. Mol. Cell Biol.* **2005**, *6*, 777–788. [CrossRef] [PubMed]
89. Okano, H.; Temple, S. Cell types to order: Temporal specification of CNS stem cells. *Curr. Opin. Neurobiol.* **2009**, *19*, 112–119. [CrossRef] [PubMed]
90. Miyata, T.; Kawaguchi, D.; Kawaguchi, A.; Gotoh, Y. Mechanisms that regulate the number of neurons during mouse neocortical development. *Curr. Opin. Neurobiol.* **2010**, *20*, 22–28. [CrossRef]
91. LoTurco, J.J.; Owens, D.F.; Heath, M.J.; Davis, M.B.; Kriegstein, A.R. GABA and glutamate depolarize cortical progenitor cells and inhibit DNA synthesis. *Neuron* **1995**, *15*, 1287–1298. [CrossRef]
92. Tochitani, S.; Sakata-Haga, H.; Fukui, Y. Embryonic exposure to ethanol disturbs regulation of mitotic spindle orientation via GABA(A) receptors in neural progenitors in ventricular zone of developing neocortex. *Neurosci. Lett.* **2010**, *472*, 128–132. [CrossRef] [PubMed]
93. Tochitani, S.; Furukawa, T.; Bando, R.; Kondo, S.; Ito, T.; Matsushima, Y.; Kojima, T.; Matsuzaki, H.; Fukuda, A. GABAA Receptors and Maternally Derived Taurine Regulate the Temporal Specification of Progenitors of Excitatory Glutamatergic Neurons in the Mouse Developing Cortex. *Cereb. Cortex* **2021**, *31*, 4554–4575. [CrossRef] [PubMed]
94. Tochitani, S.; Kondo, S. Immunoreactivity for GABA, GAD65, GAD67 and Bestrophin-1 in the meninges and the choroid plexus: Implications for non-neuronal sources for GABA in the developing mouse brain. *PLoS ONE* **2013**, *8*, e56901. [CrossRef] [PubMed]
95. Palackal, T.; Moretz, R.; Wisniewski, H.; Sturman, J. Abnormal visual cortex development in the kitten associated with maternal dietary taurine deprivation. *J. Neurosci. Res.* **1986**, *15*, 223–239. [CrossRef] [PubMed]
96. Behar, T.N.; Smith, S.V.; Kennedy, R.T.; McKenzie, J.M.; Maric, I.; Barker, J.L. GABA(B) receptors mediate motility signals for migrating embryonic cortical cells. *Cereb. Cortex* **2001**, *11*, 744–753. [CrossRef]
97. Furukawa, T.; Yamada, J.; Akita, T.; Matsushima, Y.; Yanagawa, Y.; Fukuda, A. Roles of taurine-mediated tonic GABAA receptor activation in the radial migration of neurons in the fetal mouse cerebral cortex. *Front. Cell. Neurosci.* **2014**, *8*, 88. [CrossRef]
98. Nimmervoll, B.; Denter, D.G.; Sava, I.; Kilb, W.; Luhmann, H.J. Glycine receptors influence radial migration in the embryonic mouse neocortex. *Neuroreport* **2011**, *22*, 509–513. [CrossRef]
99. Kilb, W.; Ikeda, M.; Uchida, K.; Okabe, A.; Fukuda, A.; Luhmann, H.J. Depolarizing glycine responses in Cajal-Retzius cells of neonatal rat cerebral cortex. *Neuroscience* **2002**, *112*, 299–307. [CrossRef]
100. Kirmse, K.; Dvorzhak, A.; Henneberger, C.; Grantyn, R.; Kirischuk, S. Cajal Retzius cells in the mouse neocortex receive two types of pre- and postsynaptically distinct GABAergic inputs. *J. Physiol.* **2007**, *585*, 881–895. [CrossRef]
101. Qian, T.; Chen, R.; Nakamura, M.; Furukawa, T.; Kumada, T.; Akita, T.; Kilb, W.; Luhmann, H.J.; Nakahara, D.; Fukuda, A. Activity-dependent endogenous taurine release facilitates excitatory neurotransmission in the neocortical marginal zone of neonatal rats. *Front. Cell. Neurosci.* **2014**, *8*, 33. [CrossRef] [PubMed]
102. Flint, A.C.; Liu, X.; Kriegstein, A.R. Nonsynaptic Glycine Receptor Activation during Early Neocortical Development. *Neuron* **1998**, *20*, 43–53. [CrossRef]
103. Lehtinen, M.K.; Walsh, C.A. Neurogenesis at the brain-cerebrospinal fluid interface. *Annu. Rev. Cell Dev. Biol.* **2011**, *27*, 653–679. [CrossRef] [PubMed]
104. Lehtinen, M.K.; Zappaterra, M.W.; Chen, X.; Yang, Y.J.; Hill, A.D.; Lun, M.; Maynard, T.; Gonzalez, D.; Kim, S.; Ye, P.; et al. The cerebrospinal fluid provides a proliferative niche for neural progenitor cells. *Neuron* **2011**, *69*, 893–905. [CrossRef] [PubMed]
105. You, J.S.; Chang, K.J. Taurine protects the liver against lipid peroxidation and membrane disintegration during rat hepatocarcinogenesis. *Adv. Exp. Med. Biol.* **1998**, *442*, 105–112. [CrossRef]
106. Lambert, I.H. Regulation of the cellular content of the organic osmolyte taurine in mammalian cells. *Neurochem. Res.* **2004**, *29*, 27–63. [CrossRef]
107. El Idrissi, A. Taurine increases mitochondrial buffering of calcium: Role in neuroprotection. *Amino Acids* **2008**, *34*, 321–328. [CrossRef]
108. Suzuki, T.; Suzuki, T.; Wada, T.; Saigo, K.; Watanabe, K. Taurine as a constituent of mitochondrial tRNAs: New insights into the functions of taurine and human mitochondrial diseases. *EMBO J.* **2002**, *21*, 6581–6589. [CrossRef]
109. Asano, K.; Suzuki, T.; Saito, A.; Wei, F.Y.; Ikeuchi, Y.; Numata, T.; Tanaka, R.; Yamane, Y.; Yamamoto, T.; Goto, T.; et al. Metabolic and chemical regulation of tRNA modification associated with taurine deficiency and human disease. *Nucleic. Acids Res.* **2018**, *46*, 1565–1583. [CrossRef]
110. Fakruddin, M.; Wei, F.Y.; Suzuki, T.; Asano, K.; Kaieda, T.; Omori, A.; Izumi, R.; Fujimura, A.; Kaitsuka, T.; Miyata, K.; et al. Defective Mitochondrial tRNA Taurine Modification Activates Global Proteostress and Leads to Mitochondrial Disease. *Cell Rep.* **2018**, *22*, 482–496. [CrossRef]
111. Jong, C.J.; Sandal, P.; Schaffer, S.W. The Role of Taurine in Mitochondria Health: More Than Just an Antioxidant. *Molecules* **2021**, *26*, 4913. [CrossRef] [PubMed]
112. Schaffer, S.W.; Azuma, J.; Mozaffari, M. Role of antioxidant activity of taurine in diabetes. *Can. J. Physiol. Pharmacol.* **2009**, *87*, 91–99. [CrossRef]

113. Baseggio Conrado, A.; D'Angelantonio, M.; D'Erme, M.; Pecci, L.; Fontana, M. The Interaction of Hypotaurine and Other Sulfinates with Reactive Oxygen and Nitrogen Species: A Survey of Reaction Mechanisms. *Adv. Exp. Med. Biol.* **2017**, *975 Pt 1*, 573–583. [CrossRef] [PubMed]
114. Ra, S.G.; Choi, Y.; Akazawa, N.; Ohmori, H.; Maeda, S. Taurine supplementation attenuates delayed increase in exercise-induced arterial stiffness. *Appl. Physiol. Nutr. Metab.* **2016**, *41*, 618–623. [CrossRef] [PubMed]
115. Marcinkiewicz, J.; Kontny, E. Taurine and inflammatory diseases. *Amino Acids* **2014**, *46*, 7–20. [CrossRef] [PubMed]
116. Taranukhin, A.G.; Taranukhina, E.Y.; Saransaari, P.; Djatchkova, I.M.; Pelto-Huikko, M.; Oja, S.S. Taurine reduces caspase-8 and caspase-9 expression induced by ischemia in the mouse hypothalamic nuclei. *Amino Acids* **2008**, *34*, 169–174. [CrossRef] [PubMed]
117. Murakami, S. Role of taurine in the pathogenesis of obesity. *Mol. Nutr. Food Res.* **2015**, *59*, 1353–1363. [CrossRef] [PubMed]
118. Ra, S.G.; Miyazaki, T.; Ishikura, K.; Nagayama, H.; Komine, S.; Nakata, Y.; Maeda, S.; Matsuzaki, Y.; Ohmori, H. Combined effect of branched-chain amino acids and taurine supplementation on delayed onset muscle soreness and muscle damage in high-intensity eccentric exercise. *J. Int. Soc. Sports Nutr.* **2013**, *10*, 51. [CrossRef]
119. Tochitani, S. Vertical transmission of gut microbiota: Points of action of environmental factors influencing brain development. *Neurosci. Res.* **2021**, *168*, 83–94. [CrossRef]
120. Ratsika, A.; Codagnone, M.C.; O'Mahony, S.; Stanton, C.; Cryan, J.F. Priming for Life: Early Life Nutrition and the Microbiota-Gut-Brain Axis. *Nutrients* **2021**, *13*, 426. [CrossRef]
121. Clemente, J.C.; Ursell, L.K.; Parfrey, L.W.; Knight, R. The impact of the gut microbiota on human health: An integrative view. *Cell* **2012**, *148*, 1258–1270. [CrossRef] [PubMed]
122. Tochitani, S.; Ikeno, T.; Ito, T.; Sakurai, A.; Yamauchi, T.; Matsuzaki, H. Administration of Non-Absorbable Antibiotics to Pregnant Mice to Perturb the Maternal Gut Microbiota Is Associated with Alterations in Offspring Behavior. *PLoS ONE* **2016**, *11*, e0138293. [CrossRef] [PubMed]
123. Sekirov, I.; Russell, S.L.; Antunes, L.C.; Finlay, B.B. Gut microbiota in health and disease. *Physiol. Rev.* **2010**, *90*, 859–904. [CrossRef] [PubMed]
124. Hofmann, A.F.; Hagey, L.R.; Krasowski, M.D. Bile salts of vertebrates: Structural variation and possible evolutionary significance. *J. Lipid. Res.* **2010**, *51*, 226–246. [CrossRef]
125. Selwyn, F.P.; Csanaky, I.L.; Zhang, Y.; Klaassen, C.D. Importance of Large Intestine in Regulating Bile Acids and Glucagon-Like Peptide-1 in Germ-Free Mice. *Drug Metab. Dispos.* **2015**, *43*, 1544–1556. [CrossRef]
126. Barnum, C.J.; Pace, T.W.; Hu, F.; Neigh, G.N.; Tansey, M.G. Psychological stress in adolescent and adult mice increases neuroinflammation and attenuates the response to LPS challenge. *J. Neuroinflammation* **2012**, *9*, 9. [CrossRef]
127. Sayin, S.I.; Wahlstrom, A.; Felin, J.; Jantti, S.; Marschall, H.U.; Bamberg, K.; Angelin, B.; Hyotylainen, T.; Oresic, M.; Backhed, F. Gut microbiota regulates bile acid metabolism by reducing the levels of tauro-beta-muricholic acid, a naturally occurring FXR antagonist. *Cell Metab.* **2013**, *17*, 225–235. [CrossRef]
128. Wahlstrom, A.; Sayin, S.I.; Marschall, H.U.; Backhed, F. Intestinal Crosstalk between Bile Acids and Microbiota and Its Impact on Host Metabolism. *Cell Metab.* **2016**, *24*, 41–50. [CrossRef]
129. Miyazaki, T.; Sasaki, S.I.; Toyoda, A.; Wei, F.Y.; Shirai, M.; Morishita, Y.; Ikegami, T.; Tomizawa, K.; Honda, A. Impaired bile acid metabolism with defectives of mitochondrial-tRNA taurine modification and bile acid taurine conjugation in the taurine depleted cats. *Sci. Rep.* **2020**, *10*, 4915. [CrossRef]
130. Stacy, A.; Andrade-Oliveira, V.; McCulloch, J.A.; Hild, B.; Oh, J.H.; Perez-Chaparro, P.J.; Sim, C.K.; Lim, A.I.; Link, V.M.; Enamorado, M.; et al. Infection trains the host for microbiota-enhanced resistance to pathogens. *Cell* **2021**, *184*, 615–627. [CrossRef]
131. Collard, J.M.; Sansonetti, P.; Papon, N. Taurine Makes Our Microbiota Stronger. *Trends Endocrinol. Metab.* **2021**, *32*, 259–261. [CrossRef] [PubMed]
132. Barker, D.J. The developmental origins of well-being. *Philos. Trans. R. Soc. Lond. B Biol. Sci.* **2004**, *359*, 1359–1366. [CrossRef] [PubMed]
133. Barker, D.J.P.; Osmond, C.; Winter, P.D.; Margetts, B.; Simmonds, S.J. Weight in Infancy and Death from Ischaemic Heart Disease. *Lancet* **1989**, *334*, 577–580. [CrossRef]
134. Gillman, M.W. Developmental origins of health and disease. *N. Engl. J. Med.* **2005**, *353*, 1848–1850. [CrossRef]
135. Lacagnina, S. The Developmental Origins of Health and Disease (DOHaD). *Am. J. Lifestyle Med.* **2020**, *14*, 47–50. [CrossRef] [PubMed]
136. Masztalerz-Kozubek, D.; Zielinska-Pukos, M.A.; Hamulka, J. Maternal Diet, Nutritional Status, and Birth-Related Factors Influencing Offspring's Bone Mineral Density: A Narrative Review of Observational, Cohort, and Randomized Controlled Trials. *Nutrients* **2021**, *13*, 2302. [CrossRef]
137. Armengaud, J.B.; Yzydorczyk, C.; Siddeek, B.; Peyter, A.C.; Simeoni, U. Intrauterine growth restriction: Clinical consequences on health and disease at adulthood. *Reprod. Toxicol.* **2021**, *99*, 168–176. [CrossRef]
138. Nam, H.K.; Lee, K.H. Small for gestational age and obesity: Epidemiology and general risks. *Ann. Pediatr. Endocrinol. Metab.* **2018**, *23*, 9–13. [CrossRef]
139. Surico, D.; Bordino, V.; Cantaluppi, V.; Mary, D.; Gentilli, S.; Oldani, A.; Farruggio, S.; Melluzza, C.; Raina, G.; Grossini, E. Preeclampsia and intrauterine growth restriction: Role of human umbilical cord mesenchymal stem cells-trophoblast cross-talk. *PLoS ONE* **2019**, *14*, e0218437. [CrossRef]

Article

Effects of Taurine Depletion on Body Weight and Mouse Behavior during Development

Miho Watanabe [1], Takashi Ito [2] and Atsuo Fukuda [1,*]

[1] Department of Neurophysiology, Hamamatsu University School of Medicine, Hamamatsu 431-3192, Japan; mihow@hama-med.ac.jp

[2] Department of Bioscience and Technology, Graduate School of Bioscience and Technology, Fukui Prefectural University, Fukui 910-1195, Japan; tito@fpu.ac.jp

* Correspondence: axfukuda@hama-med.ac.jp

Abstract: Taurine (2-aminoethanesulfonic acid) plays an important role in various physiological functions and is abundant in the brain and skeletal muscle. Extracellular taurine is an endogenous agonist of gamma-aminobutyric acid type A and glycine receptors. Taurine actively accumulates in cells via the taurine transporter (TauT). Adult taurine-knockout ($TauT^{-/-}$) mice exhibit lower body weights and exercise intolerance. To further examine the physiological role of taurine, we examined the effect of its depletion on mouse behavior, startle responses, muscular endurance, and body weight during development from postnatal day 0 (P0) until P60. In the elevated plus maze test, $TauT^{-/-}$ mice showed decreased anxiety-like behavior. In addition, $TauT^{-/-}$ mice did not show a startle response to startle stimuli, suggesting they have difficulty hearing. Wire-hang test revealed that muscular endurance was reduced in $TauT^{-/-}$ mice. Although a reduction of body weight was observed in $TauT^{-/-}$ mice during the developmental period, changes in body weight during 60% food restriction were similar to wild-type mice. Collectively, these results suggest that taurine has important roles in anxiety-like behavior, hearing, muscular endurance, and maintenance of body weight.

Keywords: taurine; taurine transporter; knockout mice; brain; skeletal muscle; body weight; behavior

Citation: Watanabe, M.; Ito, T.; Fukuda, A. Effects of Taurine Depletion on Body Weight and Mouse Behavior during Development. *Metabolites* **2022**, *12*, 631. https://doi.org/10.3390/metabo12070631

Academic Editors: Walter Wahli and David J. Beale

Received: 22 April 2022
Accepted: 7 July 2022
Published: 9 July 2022

1. Introduction

Taurine (2-aminoethanesulfonic acid) is one of the most ubiquitous and abundant amino acids in mammalian tissues [1,2], whereby it plays important roles in various physiological functions [3] such as mitochondrial translation [4], Ca^{2+} homeostasis [5], cell volume regulation [6], stability of membranes [7], and antioxidant effects [8]. Although the capacity to synthesize taurine is limited in most tissues [9], high levels of intracellular taurine are maintained by the taurine transporter (TauT/SLC6a6) on the membrane [10]. TauT is a Na^{+}-Cl^{-} ion-dependent transporter expressed ubiquitously in mammalian tissues [10]. The development of TauT-knockout ($TauT^{-/-}$) mice by two laboratories revealed the physiological functions of taurine [11,12]. $TauT^{-/-}$ mice exhibit shortened lifespan; cardiomyopathy; aging-dependent cardiac dysfunction; hepatitis; and renal, visual, auditory, and olfactory dysfunction [13–15]. Transcriptome microarray analysis of $TauT^{-/-}$ mice revealed that pro-fibrotic genes, such as S100 calcium-binding protein A4 (S100A4), actin alpha 2, smooth muscle (ACTA2), and connective tissue growth factor (CTGF), were increased in $TauT^{-/-}$ mice hearts [16]. Based on the transcriptome-pathway analysis, the genes involved in the "organization of extracellular matrix," such as galectin 3 (LGALS3), are enriched in old $TauT^{-/-}$ mice hearts, suggesting the contribution of these genes to fibrosis [16].

Taurine is present in high concentrations in the brain [17–19]. Moreover, taurine levels are high in the immature brain compared with the adult brain [18]. Taurine is required for the development of the central nervous system [20–22]. Our prior reports demonstrated that ambient taurine tonically activates gamma-aminobutyric acid type A

($GABA_A$) receptors and slows radial migration in the developing cerebral cortex [23]. In addition, taurine regulates the temporal specification of excitatory glutamatergic neuronal progenitors in the developing cortex [24]. In immature neurons, taurine inhibits the K^+-Cl^- cotransporter KCC2, which maintains low intracellular Cl^- levels critical for mediating fast-hyperpolarizing synaptic inhibition by $GABA_A$ and glycine receptors via the with-no-lysine (WNK) protein kinase signaling pathway [25]. However, the effects of taurine depletion on adult mouse behaviors, such as locomotor activity, anxiety-like behavior, and startle response, have not yet been reported.

Taurine also occurs in particularly high concentrations in skeletal muscle [11]. $TauT^{-/-}$ mice show a more than 98% reduction in muscle taurine content [12,26]. In addition, $TauT^{-/-}$ mice exhibit exercise intolerance in both treadmills and forced swimming tests [26,27], as well as muscular structural defects, such as myofibril derangement and the appearance of autophagic bodies [12,14]. $TauT^{-/-}$ mice exhibit accelerated skeletal muscle aging and growth, and differentiation factor 15 (GDF15), which is a member of the transforming growth factor-β superfamily, was elevated during aging in $TauT^{-/-}$ mice [28]. A previous report found that adult $TauT^{-/-}$ mice fed a normal diet had lower body weights and visceral fat, although the body weights of $TauT^{-/-}$ mice fed a diet containing 60% fat increased to a level comparable to that of wild-type (WT) mice [29]. $TauT^{-/-}$ mice showed lower fasting blood glucose and were more tolerant against glucose injection [29]. Metabolism of fatty acids was suppressed by the downregulation of genes and enzymes involved in fatty acids β-oxidation [27]. Taurine reduced obesity in WT mice fed a high-fat diet, suppressed increases in adipocyte size, body fat, and body weight [30] and improved insulin sensitivity, and increased energy expenditure and adaptive thermogenesis by elevating the browning of white adipose tissue [31]. Body weight and height of infants at birth were significantly higher in the high taurine intake pregnant women compared to infants in low taurine intake pregnant women [32]. However, changes in body weight of $TauT^{-/-}$ mice during development remain unclear.

In this study, to further examine the physiological role of taurine, we used $TauT^{-/-}$ mice to investigate the effects of taurine deficiency on body weight changes during development and in adult mice during food restriction. In addition, we measured muscular endurance. Furthermore, to examine the effects of taurine deficiency on behavior, we performed several behavioral analyses.

2. Results

2.1. $TauT^{-/-}$ Mice Showed Decreased Anxiety-Like Behavior

We performed a behavioral analysis of $TauT^{-/-}$ mice. To assay general locomotor activity levels and anxiety, we performed an open field test. Percentages of time spent in the center (WT, 9.3 ± 3.1%, n = 12; $TauT^{+/-}$, 7.4 ± 2.0%, n = 12; $TauT^{-/-}$, 4.7 ± 1.4%, n = 8) and time spent in the corner (WT, 59.2 ± 5.7%; $TauT^{+/-}$, 61.9 ± 3.7%; $TauT^{-/-}$, 60.4 ± 4.7%) were similar between genotypes (Figure 1A,B). Moreover, no significant difference in the total distance moved was found between genotypes (WT, 3354.1 ± 234.1 cm; $TauT^{+/-}$, 3690.5 ± 214.9 cm; $TauT^{-/-}$ mice, 3891.4 ± 582.9 cm; Figure 1C). From these results, $TauT^{+/-}$ and $TauT^{-/-}$ mice both showed normal locomotor activity. To examine anxiety-like behavior, we performed an elevated plus maze test. The percentage of time spent in the open arm was increased in $TauT^{-/-}$ mice compared with WT and $TauT^{+/-}$ mice, although there were no significant differences between groups (WT, 13.5 ± 2.3%, n = 13; $TauT^{+/-}$, 14.7 ± 4.4%, n = 11; $TauT^{-/-}$, 28.6 ± 10.5%, n = 11; Figure 1D). The percentage of time spent in the closed arm was significantly decreased in $TauT^{-/-}$ mice compared with WT and $TauT^{+/-}$ mice (WT, 78.4 ± 3.3%; $TauT^{+/-}$, 77.8 ± 5.8%; $TauT^{-/-}$, 52.8 ± 9.0%). In addition, the percentage of time spent in the center was significantly increased in $TauT^{-/-}$ mice compared with WT mice and showed an increased trend compared with WT mice (p = 0.06; WT, 8.1 ± 1.4%; $TauT^{+/-}$, 7.4 ± 1.6%; $TauT^{-/-}$, 18.6 ± 3.1%). No significant difference in total distance moved was observed between genotypes (WT, 1329.7 ± 58.2 cm; $TauT^{+/-}$, 1274.9 ± 61.1 cm; $TauT^{-/-}$,

1117.1 ± 67.0 cm; Figure 1E). These results suggest that $TauT^{-/-}$ mice showed decreased anxiety-like behavior and had difficulty making decisions underlying approach/avoidance conflict and risk assessment.

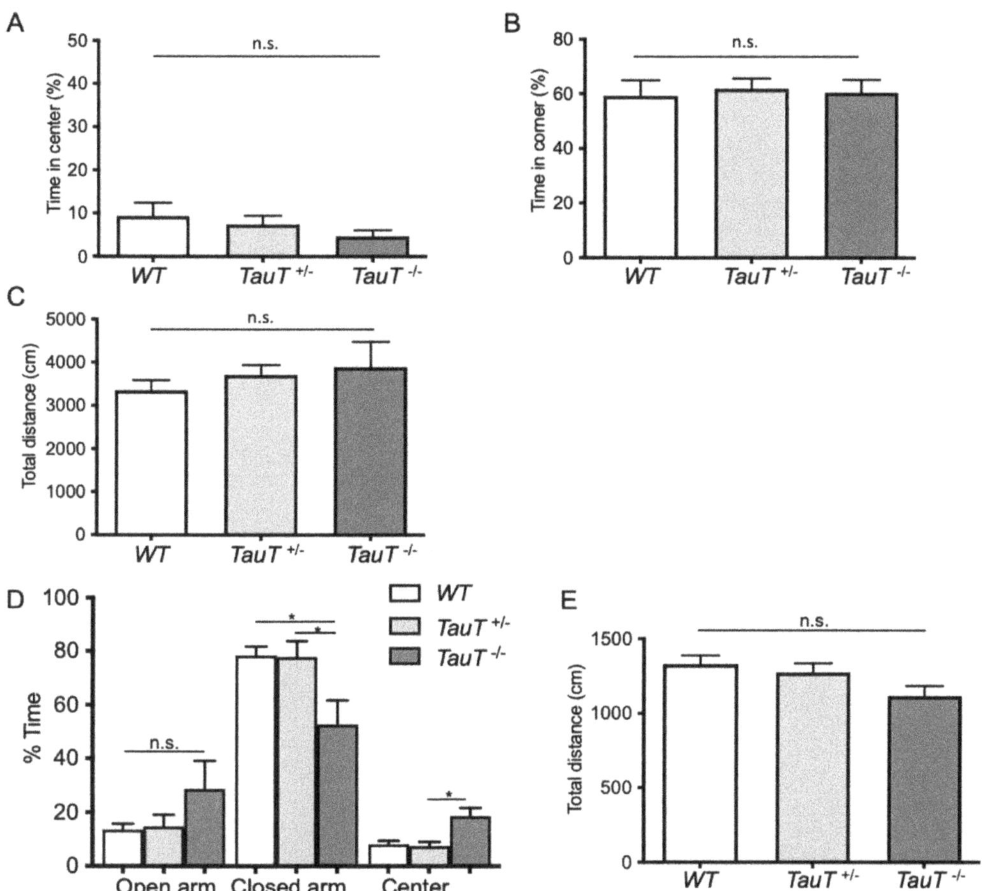

Figure 1. (**A–C**) In the open field test, mice were placed in the center of the open field apparatus and their movement was tracked for 10 min. The percentage of time spent in the center (**A**), percentage of time spent in the corner (**B**), and total distance moved in an open field (**C**) were measured for WT, $TauT^{+/-}$, and $TauT^{-/-}$ mice. (**D**,**E**) In the elevated plus maze test, mice were placed in the center of the elevated plus maze and allowed to explore for 5 min. The percentage of time spent in the open arms, closed arms, and center (* $p < 0.05$, n.s. = not significant by Kruskal–Wallis test) (**D**) and total distance moved (**E**) in the elevated plus maze were measured. Error bars represent S.E.M.

2.2. $TauT^{-/-}$ Mice Showed Difficulty Hearing

Next, we performed the startle response test and checked the dependence of startle responses on startle stimulus intensity (60 to 120-dB pulse, WT $n = 6$, $TauT^{-/-}$ $n = 6$). WT mice showed a startle response to startle pulses between 90 and 120 dB (Table 1). In total, 50% of WT mice showed a startle response to a startle pulse of 90 dB. All WT mice showed startle responses to startle pulses between 100 and 120 dB. However, no $TauT^{-/-}$ mice exhibited a startle response to any startle pulse. From these results, $TauT^{-/-}$ mice likely have difficulty hearing.

Table 1. Startle response test. Numbers of mice showing a startle response to the startle pulse.

Gene Type	No. of Mice Showed Startle Response (%)						
	60 dB	**70 dB**	**80 dB**	**90 dB**	**100 dB**	**110 dB**	**120 dB**
WT	0/0(0%)	0/0(0%)	0/0(0%)	3/6(50%)	6/6(100%)	6/6(100%)	6/6(100%)
TauT$^{-/-}$	0/0(0%)	0/0(0%)	0/0(0%)	0/0(0%)	0/0(0%)	0/0(0%)	0/0(0%)

2.3. *Muscular Endurance Was Reduced in TauT*$^{-/-}$ *Mice*

A previous report suggested that exercise capacity (as determined by a treadmill test) was reduced in *TauT*$^{-/-}$ mice [27]. Therefore, to measure muscular endurance, a wire-hang test was performed. Each mouse was placed on a wire mesh and then turned upside down. The latency to fall was measured. Latency to fall was reduced in *TauT*$^{-/-}$ mice compared with WT mice (WT, 552.6 $\pm$ 32.4 s, n = 7; *TauT*$^{-/-}$, 462.6 $\pm$ 32.6 s, n = 6; Figure 2). These results suggest that muscular endurance was reduced in *TauT*$^{-/-}$ mice.

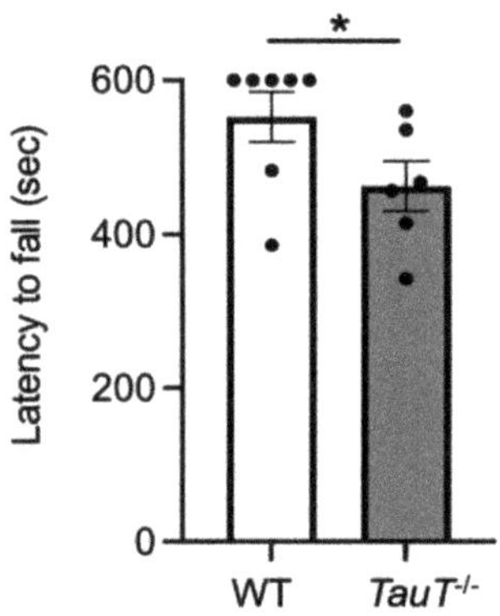

Figure 2. Muscular endurance was reduced in *TauT*$^{-/-}$ mice. Wire-hang test was performed. Average latency to fall from the wire mesh of two trials in WT and *TauT*$^{-/-}$ mice. * $p < 0.05$ by Mann-Whitney test. Error bars represent S.E.M. Closed circles represent animals.

2.4. *Body Weight Gain Was Reduced in TauT*$^{-/-}$ *Mice during Development*

Adult *TauT*$^{-/-}$ mice showed reduced body weights [29], so we examined changes in body weight during development from postnatal day 0 (P0) until P60. The body weight of each mouse was measured every 2 days. At birth, there were no differences in body weights between genotypes for either male (WT, 1.39 $\pm$ 0.05 g, n = 15; *TauT*$^{+/-}$, 1.41 $\pm$ 0.03 g, n = 39; *TauT*$^{-/-}$, 1.35 $\pm$ 0.02 g, n = 13) or female (WT, 1.37 $\pm$ 0.03 g, n = 23; *TauT*$^{+/-}$, 1.38 $\pm$ 0.02 g, n = 39; *TauT*$^{-/-}$, 1.38 $\pm$ 0.03 g, n = 16) mice (Figure 3A,B). However, body weights of male *TauT*$^{-/-}$ mice were significantly lower compared with male WT and *TauT*$^{+/-}$ mice between P4 onward (WT, n = 12; *TauT*$^{+/-}$, n = 32; *TauT*$^{-/-}$, n = 12) (Figure 3C). In addition, a reduction in body weights of female *TauT*$^{-/-}$ mice was observed between P10 onward (WT, n = 16; *TauT*$^{+/-}$, n = 32; *TauT*$^{-/-}$, n = 15) (Figure 3D). The difference in body weight seems to become apparent around P20 when mice start self-feeding. Notably, the difference in body weights between WT and *TauT*$^{-/-}$ mice was greater in male mice than in female mice.

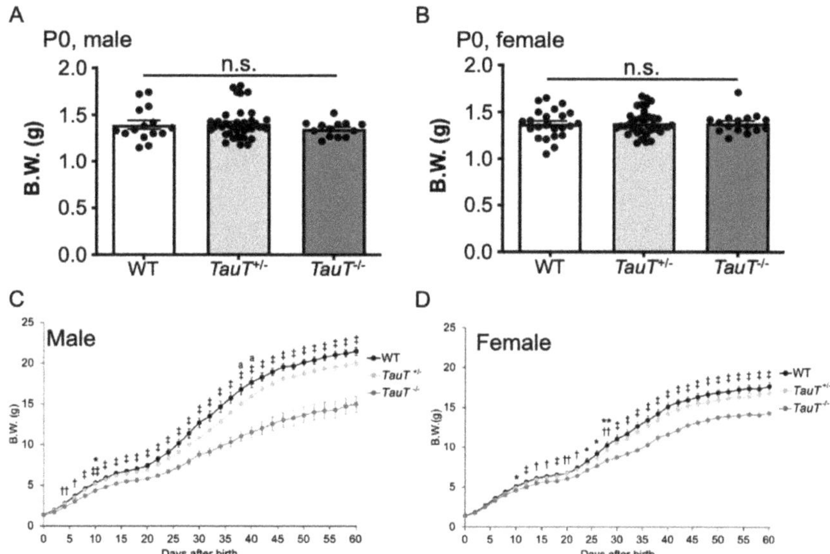

Figure 3. *TauT*$^{-/-}$ mice exhibited lower body weights during development. (**A**) Body weights of male WT, *TauT*$^{+/-}$, and *TauT*$^{-/-}$ pups at P0. Closed circles represent animals. (**B**) Body weights of female WT, *TauT*$^{+/-}$, and *TauT*$^{-/-}$ pups at P0. (**C**) Changes in body weights of male WT, *TauT*$^{+/-}$, and *TauT*$^{-/-}$ mice during development. (**D**) Changes in body weights of female WT, *TauT*$^{+/-}$, and *TauT*$^{-/-}$ mice during development. Error bars represent S.E.M. * $p < 0.05$ *TauT*$^{-/-}$ vs. WT, ** $p < 0.01$ *TauT*$^{-/-}$ vs. WT, † $p < 0.05$ *TauT*$^{-/-}$ vs. WT and *TauT*$^{+/-}$, ‡ $p < 0.01$ *TauT*$^{-/-}$ vs. WT and *TauT*$^{+/-}$, †† $p < 0.05$ *TauT*$^{-/-}$ vs. *TauT*$^{+/-}$, ‡‡ $p < 0.01$ *TauT*$^{-/-}$ vs. *TauT*$^{+/-}$, and ^{a} $p < 0.05$ *TauT*$^{+/-}$ vs. WT by Kruskal-Wallis test.

2.5. Body Weight Changes during 60% Food Restriction Were Similar in WT and TauT$^{-/-}$ Mice

A previous report showed that the weight of visceral fat was significantly less in *TauT*$^{-/-}$ mice compared with WT mice; however, when fed a high-fat diet, *TauT*$^{-/-}$ mice exhibited body weights comparable to WT mice and dramatically increased abdominal fat after 16 weeks [29]. Therefore, we examined the effect of a 60% food restriction. Mice were given a food pellet equal to 60% of daily food intake before the onset of the dark phase for 2 weeks. Subsequently, mice were given food *ad libitum* for 2 weeks. Body weights were measured every day. Body weights were markedly decreased during the first 5 days after starting 60% food restriction, then gradually decreased in both WT and *TauT*$^{-/-}$ mice (WT, $n = 7$; *TauT*$^{-/-}$, $n = 7$; Figure 4A,B). After WT and *TauT*$^{-/-}$ mice were given food *ad libitum*; their body weights increased in both WT and *TauT*$^{-/-}$ mice. Indeed, the remarkable weight loss observed during the first 5 days after starting 60% food restriction gradually recovered in both WT and *TauT*$^{-/-}$ mice (Figure 4C). On the first day, mice were given food *ad libitum*; the weight gain was significantly lower in *TauT*$^{-/-}$ mice (2.9 $\pm$ 0.1 g) compared with WT mice (4.2 $\pm$ 0.2 g). Thus, *TauT*$^{-/-}$ mice showed an apparent disturbance of body weight recovery after food restriction compared with WT mice, although their body weight reduction during food restriction was comparable.

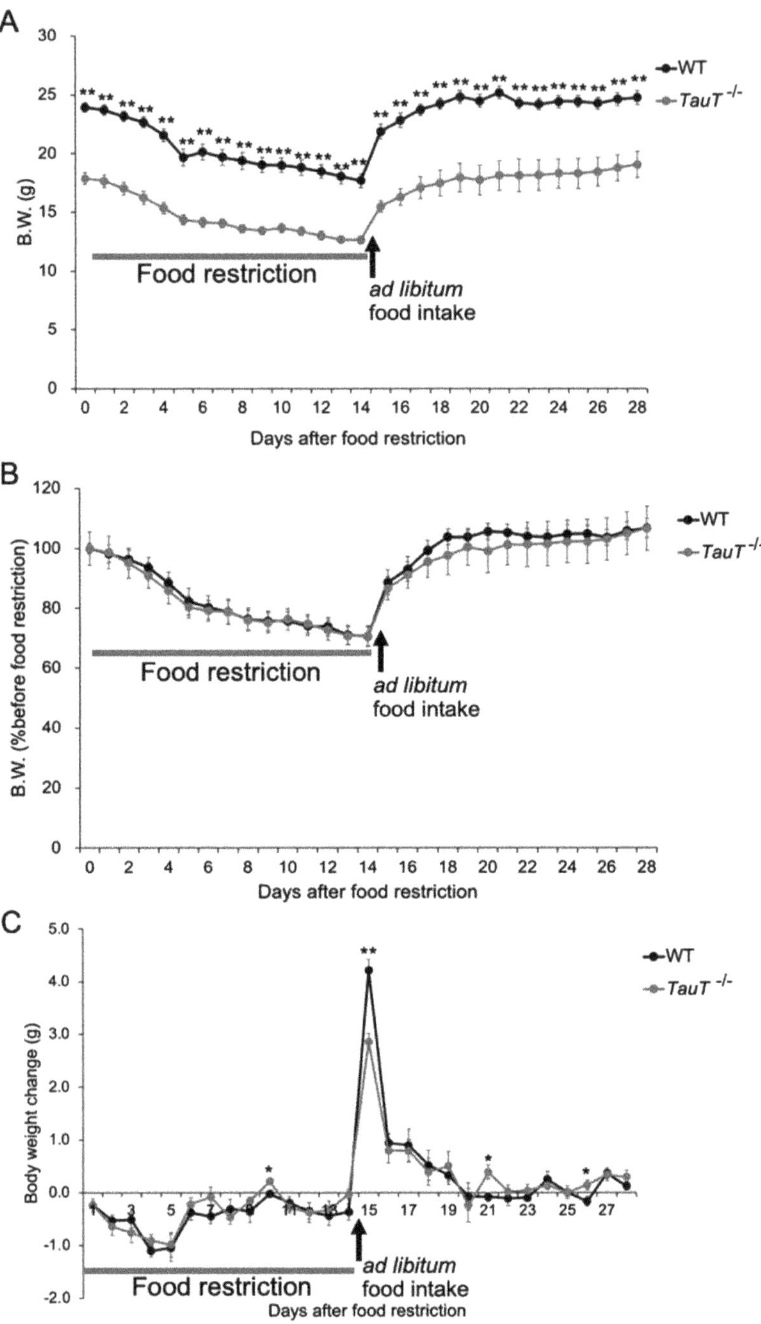

Figure 4. Changes in body weights following 60% food restriction in WT and $TauT^{-/-}$ mice. (**A**) Changes in body weight after 60% food restriction and recovery (food *ad libitum*) in WT mice and $TauT^{-/-}$ mice. (**B**) Body weight changes are shown as a percentage of the initial body weight before 60% food restriction. (**C**) Body weight loss or gain after 60% food restriction and recovery (food *ad libitum*) in WT mice and $TauT^{-/-}$ mice. Error bars represent S.E.M. * $p < 0.05$, ** $p < 0.01$ by Kruskal-Wallis test.

2.6. Comprehensive Analysis of Protein Kinases and Target Proteins

Metabolic processes are commonly regulated by a series of phosphorylation events by kinases. Therefore, we performed an active site-directed competition binding assay to quantitatively measure interactions between kinases and target proteins to compare differences between WT and $TauT^{-/-}$ mice. Nineteen proteins showed significant changes in expression levels between WT and $TauT^{-/-}$ mice. Moreover, three proteins exhibited significant changes in the phosphorylation of analyzed sites (Table S1). The most prominent protein alteration was phosphorylation of signal transducer and activator of transcription 3 (STAT3) at tyrosine 705, which was increased by 9449% in the brains of $TauT^{-/-}$ mice compared with WT mice.

3. Discussion

We have shown that taurine deficiency in mice decreased anxiety-like behaviors, abolished startle responses, and reduced muscular endurance. Furthermore, taurine deficiency caused lower body weight gain during development. Our findings demonstrate the importance of taurine in brain function, muscular endurance, and energy metabolism.

Our results show that $TauT^{-/-}$ mice exhibited decreased anxiety-like behaviors, as indicated by reduced time in the closed arm of the elevated plus maze test and increased time spent in the center square. Time spent exploring the center square has been suggested to reflect decision-making underlying approach-avoidance conflict and risk assessment. Although $TauT^{-/-}$ mice exhibit visual dysfunction due to retinal degeneration [11] and hearing loss [13], the elevated plus maze test is based on mouse exploratory patterns that avoid open spaces and are motivated by thigmotaxis; the tendency to be close to vertical surfaces is far more reliant on tactile inputs compared with vision [33]. Therefore, we consider that visual dysfunction and hearing loss have little effect on our behavioral tests. Taurine is present at high concentrations in the brain [11], whereby it acts as a partial agonist of $GABA_A$ [34] and glycine receptors [35]. Although the physiological roles of taurine in the brain are still not fully understood, taurine is required for normal nervous system development [22]. In the developing brain, the taurine concentration is 3–4 times higher than present in the adult brain [18]. After the first postnatal week, taurine concentrations are downregulated [18]. Kittens from taurine-deficient mothers exhibit smaller brain weights and abnormal morphologies in the visual cortex [36] and cerebellum [37]. Ambient taurine in the mouse developing fetal cerebral cortex tonically activates $GABA_A$ receptors and slows radial migration [23]. In addition, taurine increases the proliferation of neural stem cells [24,38]. Taurine modulates the switch of GABA responses from depolarizing to hyperpolarizing by suppressing KCC2 in the embryonic brain via WNK kinase signaling [25]. In the adult brain, taurine also activates extrasynaptic $GABA_A$ receptors in the thalamus [39]. $TauT^{-/-}$ mice show impaired GABAergic inhibition in the striatum [40]. Therefore, taurine deficiency in the brain may cause structural deficits or an imbalance in inhibitory and excitatory neurotransmission in neuronal circuits controlling anxiety behavior, thereby causing abnormal behavior in $TauT^{-/-}$ mice.

$TauT^{-/-}$ mice did not show startle responses to startle stimuli. Taurine is present at high concentrations in the organ of Corti [41] and inferior colliculus [42]. Since taurine concentrations are relatively high in glia [43], *TauT* knockout in glia cells may exacerbate degeneration of auditory nerve fibers, resulting in hearing loss [13]. Six-month-old $TauT^{-/-}$ mice reportedly show 20-dB higher auditory brainstem response thresholds due to accelerated losses of outer and inner hair cells [13,44]. Taurine reduces neuronal activity via glycine receptors in the rat inferior colliculus [45]. In addition, taurine serves as a neuromodulator to enhance GABAergic and glycinergic transmission in the rat anteroventral cochlear nucleus [46]. Therefore, taurine may regulate neurotransmission of the central auditory system and its deficiency appears to cause difficulty hearing. The auditory brainstem response should be measured to precisely show $TauT^{-/-}$ mice exhibit hearing loss. Further experiment is needed at this point.

TauT$^{-/-}$ mice exhibited reduced body weights during the developmental period, although no difference was observed at birth. *TauT*$^{-/-}$ mice reportedly show lower body weights and less abdominal fat at 3 months, although both food and water intake were unchanged [12,29]. *TauT*$^{-/-}$ mice also exhibit lower weights of heart and tibial anterior muscle [12], as well as lower fasting blood glucose levels and tolerance against glucose injection [29]. Skeletal muscle lactate content is higher in *TauT*$^{-/-}$ mice than in WT mice [15], suggesting that the glycolytic pathway may be accelerated in *TauT*$^{-/-}$ mice. Acceleration of glycolysis can enhance glucose disposal from blood. In addition, several genes involved in the oxidation of fatty acids are downregulated in *TauT*$^{-/-}$ mice [15], suggesting that their fatty acid utilization is reduced. Accordingly, glucose disposal may be enhanced in *TauT*$^{-/-}$ mice as a compensatory mechanism. Therefore, imbalances in energy production and expenditure likely cause a reduction in body weight during development. Plasma insulin levels were lower in *TauT*$^{-/-}$ mice compared with WT mice due to decreased numbers of beta cells in the pancreas [29]. Insulin promotes the conversion of glucose into fat. Therefore, lower insulin levels may reduce the conversion of glucose into fat, which may cause reductions in fat and body weight in *TauT*$^{-/-}$ mice during development. In our comprehensive analysis of protein kinases and target proteins, STAT3 phosphorylation was significantly increased in *TauT*$^{-/-}$ mouse brains. Leptin, a hormone synthesized and released from adipose tissue, plays a central role in the long-term control of body weight, acting mainly in the central nervous system to induce satiety, energy expenditure, and glycolysis through the STAT3 pathway [47,48]. Leptin binding to the long isoform of leptin receptor (LepRb) results in its dimerization, leading to the formation of the LepRb/Janus kinase 2 (JAK2) complex. The activated JAK2 phosphorylates itself and also Tyr985, Tyr1077, and Tyr1138 in LepRb. STAT3 binds to phospho-Tyr1077 in LepRb and are subsequently phosphorylated. Active STAT3 dimers then translocate to the nucleus and activate the transcription of their target genes, which mediate leptin's anorexigenic effect. Therefore, there is a possibility that enhanced leptin signaling causes body weight and fat reduction in *TauT*$^{-/-}$ mice during development. Taurine depletion did not affect body weight loss during 60% food restriction. Significant body weight reduction in *TauT*$^{-/-}$ mice compared to WT mice was still observed during the food restriction and recovery period. *TauT*$^{-/-}$ mice exhibited a significant reduction in body weight gain compared to WT mice on the first day of *ad libitum* food after food restriction. Phosphorylated STAT3 activates anorexigenic neurons that express proopiomelanocortin while inhibiting orexigenic neurons that express Agouti-related protein, thereby inhibiting food intake and increasing energy expenditure [47]. Combined with a previous report of *TauT*$^{-/-}$ mice showing reduced abdominal fat weights [29], the present results show that extremely high levels of STAT3 phosphorylation may activate leptin's anorexigenic effect, thereby reducing food intake in *TauT*$^{-/-}$ mice even on the first day of *ad libitum* food after food restriction and disturb body weight recovery. The difference in body weights between WT and *TauT*$^{-/-}$ mice was greater in male mice compared with female mice. The mechanism of these sex differences is unclear, but a previous paper showed that taurine supplementation increased the level of testosterone in male rats [49], that increased skeletal muscle. Therefore, there is a possibility that taurine depletion may reduce testosterone levels, which leads to a reduction of muscle mass, and results in body weight loss, especially in male mice. Further studies are needed to show the sex differences in weight regulation by taurine.

Taurine contents in skeletal muscle are high [11,12]. *TauT*$^{-/-}$ mice reportedly show reduced swimming endurance times [12], as well as decreased maximum running speed and endurance duration in treadmill running tests [26,27], suggesting they exhibit exercise intolerance. Notably, taurine supplementation prolonged the time to exhaustion during treadmill running [50]. *TauT*$^{-/-}$ mice show decreased myofiber size and myofiber disruption [12]. We previously performed LC-MS-based metabolome analysis and suggested that organic osmolytes such as betaine, glycerophosphocholine, and amino acids were increased in the heart muscle of *TauT*$^{-/-}$ mice by compensating mechanisms for taurine depletion [51], although the expression of related gene expressions, such as betaine/GABA transporter-1 (BGT-1; Slc6a12), carnitine transporter Slc22a4 and Slc22a5 are not changed [52]. The mRNA expressions of heat shock protein 700 (Hsp70), the amino acid transporters (Slc38a2; ATA2), DnaJ (Hsp40) homolog, subfamily B, member 1 (Dnajb1), tumor necrosis factor receptor superfamily, member 12a (Tnfrsf12a), and S100 calcium-binding proteins (S100A4, S100A9), which are upregulated by osmotic stress, are elevated in both heart and skeletal muscle of the $TauT^{-/-}$ mice [15], suggesting that dysfunction of osmoregulatory system impairs the tibial anterior muscle volume regulation. Taurine depletion reduced muscular endurance in this study, supporting previous findings of an important role for taurine in the maintenance of skeletal muscle function. Furthermore, STAT3 signaling plays important roles in regulating skeletal muscle mass, repair, and diseases [53]. Chronic activation of the interleukin 6/JAK/STAT3 signaling pathway is involved in muscle wasting. Activated STAT3 promotes skeletal muscle atrophy in muscle diseases, such as muscular dystrophy, cancer, and sepsis [54,55]. Therefore, very high levels of STAT3 phosphorylation may cause skeletal muscle abnormalities, thereby decreasing muscle endurance in $TauT^{-/-}$ mice.

The limitation of this study is that we used whole-body *TauT* knockout mice. Since taurine is ubiquitous and is abundant free amino acid in the heart, retina, skeletal muscle, brain, and leukocytes [1,2], whole body *TauT* knockout causes various effects. Therefore, to demonstrate the functional role of taurine in specific tissue precisely, we need to generate tissue-specific *TauT* knockout mice. If we knockout *TauT* in a specific region of the brain, we can demonstrate the role of taurine in the behavior, excluding the effects of vision dysfunction and hearing loss. If we generate the time-specific *TauT* knockout mice, we can demonstrate the role of taurine at specific time points during development and in the adult mice. Further studies are needed to determine the precise role of taurine using cell/tissue- and time-specific *TauT* knockout mice. We have shown the effects of taurine depletion on anxiety-like behavior, startle response, muscular endurance, and body weight during development. However, the detailed mechanism by which taurine depletion causes such effects remains unclear. We have shown that $TauT^{-/-}$ mice did not show startle responses. This result suggests that $TauT^{-/-}$ mice might have hearing difficulty. However, we need to perform the auditory brainstem response test to confirm hearing difficulty in $TauT^{-/-}$ mice and to locate hearing deficits along the auditory nerve pathway. Because $TauT^{-/-}$ mice showed degeneration in the inner ear and auditory nerve, it is speculated that taurine has a role in the regulating survival of cells related to auditory function. Since taurine also acts as a partial agonist of $GABA_A$ and glycine receptors, taurine might have a role in modulating the neurotransmission in the auditory nerve tract. Further studies are needed to show the underlying mechanisms of our results.

In conclusion, the present study suggests that taurine plays important roles in the maintenance of mouse behavior, hearing, body weight, and skeletal muscle function.

4. Materials and Methods

4.1. Animals

TauT-KO and littermate mice (C57BL/6 background) were used. The generation of TauT KO-mice was previously described [12]. Heterozygous TauT-KO male and female mice were mated, and homozygous TauT-KO ($TauT^{-/-}$, male n = 51, female n = 17), heterozygous TauT-KO ($TauT^{+/-}$, male n = 62, female n = 39), and wild-type (WT, male n = 60, female n = 24) mice were used for experiments. Genotyping for TauT-KO mice was carried out by PCR of mouse tail DNA using the following primers: 5′-GGTGTCACTAAGACGTGAGTTG-3′, 5′-CAGACAGCACAAAGTCGATCT-3′, and 5′-TGCTAAAGCGCATGCTCCAGACTG-3′. Mice were kept in a 12-h light/dark cycle (lights on at 07:00 a.m.) with access to food and water *ad libitum* unless otherwise noted. All experimental procedures were approved by the institutional animal care and use committee of Hamamatsu University School of Medicine. With regard to the ethical use of animals for experimentation, all efforts were made to minimize the number of animals used and their suffering.

4.2. Measurements of Body Weight and Food Restriction

Body weights of male and female mice aged P0–P60 were measured once every 2 days between 16:00 and 19:00 h. For the food-restriction experiment, daily food consumption of WT mice was measured for 10 days and average food consumption per gram of body weight was calculated. After 7 days of individual housing for acclimation, mice were given a food pellet equal to 60% of their average daily food intake. Body weights were measured between 16:00 and 19:00 h for 14 days during 60% food restriction. After food restriction, mice were allowed free access to food, and body weights were measured for an additional 14 days to observe recovery.

4.3. Behavioral Analysis

For the open field test, the size of the open field box was 42 × 42 × 40 cm. Mice were placed in the center of the chamber at the beginning of each test. The movement of each mouse was tracked for 10 min. Locomotor activity was recorded and analyzed using a video tracking system (SMART 3.0; Panlab, Cornellà de Llobregat, Spain). For the elevated plus maze test, we used an elevated plus maze apparatus comprising a central part (6 × 6 cm), two opposing open arms (40 × 6 cm), and two opposing closed arms (40 × 6 × 15 cm). The apparatus was raised 40 cm above the floor. Mice were placed in the center of the apparatus facing the closed arm and allowed to explore the maze for 5 min freely. Time spent and distance traveled in the center, open arms, and closed arms were measured using a video tracking system. For the acoustic startle response test, a startle box system (Panlab, Cornellà de Llobregat, Spain) containing a sound generator and accelerometer was used to record the amplitude of each mouse's startle response. Mice were placed into a plexiglass cylinder and set on a weight-transducing platform in the sound-attenuating chamber. A white background noise (70 dB) was generated throughout the experiment. The acoustic startle response test began with 5 min of acclimatization. Mice were randomly exposed to startle stimuli (from 60 to 120 dB, 10-dB increments, 40-ms duration) with a variable (10–20 s) inter-stimulus interval to avoid habituation. Each sound intensity was presented six times. The peak amplitude of the startle response was measured.

4.4. Kinexus Protein Microarray

Brains from WT and $TauT^{-/-}$ mice were cut into small pieces and rinsed in ice-cold phosphate-buffered saline to remove the blood and other contamination on the surface. Next, brains were wrapped in tinfoil and snap-frozen in liquid nitrogen. Subsequently, brains were shipped in dry ice to Kinexus Bioinformatics Corporation (Vancouver, BC, Canada) for Kinexus KAM-880 Antibody Microarray, which includes 518 pan-specific (for expression levels of these phosphoproteins) and 359 phosphorylation-site-specific antibodies (for phosphorylation). Briefly, 50 g of cell lysate from each sample was covalently labeled with a fluorescent dye. Free dye molecules were then removed at the completion of labeling reactions by gel filtration. After blocking non-specific binding sites, an incubation chamber was mounted onto the microarray to permit the loading of WT and $TauT^{-/-}$ samples side by side on the same chip and to prevent mixing of the samples. Following sample incubation, unbound proteins were washed away. Each array produced a pair of 16-bit images, which were captured with a Perkin-Elmer ScanArray Reader laser array scanner (Waltham, MA, USA). Signal quantification was performed with ImaGene 8.0 from BioDiscovery (El Segundo, CA, USA), using the predetermined settings for spot segmentation and background correction. The resultant changes in spot intensity are expressed as the percentages of change for $TauT^{-/-}$ mice compared with WT mice [% change from control (% CFC) = [(globally normalized signal intensity of $TauT^{-/-}$ sample /globally normalized signal intensity of WT sample) × 100] − 100]. Changes of 50% CFC were considered significant changes. Globally normalized signal intensity is the background-corrected intensity value.

4.5. Statistical Analysis

Data are presented as mean ± standard error of the mean (SEM). Comparisons between the two groups were made using the Mann–Whitney test. Multiple comparisons were made using Kruskal–Wallis one-way ANOVA followed by Dunn's multiple comparison test. Statistical significance was defined as $p < 0.05$. All statistical analyses were performed using GraphPad Prism (version 9.2.0; GraphPad Software, San Diego, CA, USA).

Supplementary Materials: The following supporting information can be downloaded at: https://www.mdpi.com/article/10.3390/metabo12070631/s1, Table S1: Effects of taurine depletion on protein expression and phosphorylation by protein microarray analysis.

Author Contributions: Conceptualization, M.W. and A.F.; methodology, M.W., T.I., and A.F.; validation, M.W. and A.F.; investigation, M.W.; data curation, M.W.; writing–original draft preparation, M.W. and A.F.; writing—review and editing, M.W. and A.F.; visualization, M.W. and A.F.; supervision, A.F.; project administration, A.F.; funding acquisition, A.F. All authors have read and agreed to the published version of the manuscript.

Funding: This research was supported by Grants-in-Aid for Scientific Research (B) (#21H02661 to A.F.), Grants-in-Aid for Transformative Research Area "Inducing lifelong plasticity (iPlasticity) by brain rejuvenation: elucidation and manipulation of critical period mechanisms" (#21H05687 to A.F.), Smoking Research Foundation (to A.F.), The Salt Science Research Foundation (to A.F.), Grants-in-Aid for Scientific Research (C) (#20K06924 to M.W.), and The Japan Epilepsy Research Foundation (to M.W.).

Institutional Review Board Statement: The animal study protocol was approved by the Institutional Review Board of Hamamatsu University School of Medicine (protocol code 2-53 and 2021.4.27).

Informed Consent Statement: Not applicable.

Data Availability Statement: Data are contained within the article.

Acknowledgments: We thank Edanz (https://jp.edanz.com/ac) for editing a draft of this manuscript (accessed on 21 April 2022).

Conflicts of Interest: The authors declare no conflict of interest.

References

1. Chesney, R.W. Taurine: Its biological role and clinical implications. *Adv. Pediatr.* **1985**, *32*, 1–42. [PubMed]
2. Chapman, R.A.; Suleiman, M.S.; Earm, Y.E. Taurine and the heart. *Cardiovasc. Res.* **1993**, *27*, 358–363. [CrossRef] [PubMed]
3. Huxtable, R.J. Actions of taurine. *Physiol. Rev.* **1992**, *72*, 101–163. [CrossRef] [PubMed]
4. Suzuki, T.; Suzuki, T.; Wada, T.; Saigo, K.; Watanabe, K. Taurine as a constituent of mitochondrial tRNAs: New insights into the functions of taurine and human mitochondrial diseases. *EMBO J.* **2002**, *21*, 6581–6589. [CrossRef] [PubMed]
5. Chen, W.Q.; Jin, H.; Nguyen, M.; Carr, J.; Lee, Y.J.; Hsu, C.C.; Faiman, M.D.; Schloss, J.V.; Wu, J.Y. Role of taurine in regulation of intracellular calcium level and neuroprotective function in cultured neurons. *J. Neurosci. Res.* **2001**, *66*, 612–619. [CrossRef]
6. Solís, J.M.; Herranz, A.S.; Herreras, O.; Lerma, J.; del Río, R.M. Does taurine act as an osmoregulatory substance in the rat brain? *Neurosci. Lett.* **1988**, *91*, 53–58. [CrossRef]
7. You, J.S.; Chang, K.J. Taurine protects the liver against lipid peroxidation and membrane disintegration during rat hepatocarcinogenesis. *Adv. Exp. Med. Biol.* **1998**, *442*, 105–112.
8. Martincigh, B.S.; Mundoma, C.; Simoyi, R.H. Antioxidant chemistry: Hypotaurine−taurine oxidation by chlorite. *J. Phys. Chem. A* **1998**, *102*, 9838–9846. [CrossRef]
9. Ide, T.; Kushiro, M.; Takahashi, Y.; Shinohara, K.; Cha, S. mRNA expression of enzymes involved in taurine biosynthesis in rat adipose tissues. *Metabolism* **2002**, *51*, 1191–1197. [CrossRef]
10. Uchida, S.; Kwon, H.M.; Yamauchi, A.; Preston, A.S.; Marumo, F.; Handler, J.S. Molecular cloning of the cDNA for an MDCK cell Na^+- and Cl^--dependent taurine transporter that is regulated by hypertonicity. *Proc. Natl. Acad. Sci. USA* **1992**, *89*, 8230–8234. [CrossRef]
11. Heller-Stilb, B.; van Roeyen, C.; Rascher, K.; Hartwig, H.G.; Huth, A.; Seeliger, M.W.; Warskulat, U.; Häussinger, D. Disruption of the taurine transporter gene (taut) leads to retinal degeneration in mice. *FASEB J.* **2002**, *16*, 231–233. [CrossRef] [PubMed]
12. Ito, T.; Kimura, Y.; Uozumi, Y.; Takai, M.; Muraoka, S.; Matsuda, T.; Ueki, K.; Yoshiyama, M.; Ikawa, M.; Okabe, M.; et al. Taurine depletion caused by knocking out the taurine transporter gene leads to cardiomyopathy with cardiac atrophy. *J. Mol. Cell. Cardiol.* **2008**, *44*, 927–937. [CrossRef] [PubMed]

13. Warskulat, U.; Heller-Stilb, B.; Oermann, E.; Zilles, K.; Haas, H.; Lang, F.; Häussinger, D. Phenotype of the taurine transporter knockout Mouse. *Methods Enzymol.* **2007**, *428*, 439–458. [CrossRef] [PubMed]
14. Ito, T.; Oishi, S.; Takai, M.; Kimura, Y.; Uozumi, Y.; Fujio, Y.; Schaffer, S.W.; Azuma, J. Cardiac and skeletal muscle abnormality in taurine transporter-knockout mice. *J. Biomed. Sci.* **2010**, *17*, S20. [CrossRef]
15. Ito, T.; Yoshikawa, N.; Inui, T.; Miyazaki, N.; Schaffer, S.W.; Azuma, J. Tissue depletion of taurine accelerates skeletal muscle senescence and leads to early death in mice. *PLoS ONE* **2014**, *9*, e107409. [CrossRef] [PubMed]
16. Ito, T.; Hanahata, Y.; Kine, K.; Murakami, S.; Schaffer, S.W. Tissue taurine depletion induces profibrotic pattern of gene expression and causes aging-related cardiac fibrosis in heart in mice. *Biol. Pharm. Bull.* **2018**, *41*, 1561–1566. [CrossRef]
17. Agrawal, H.C.; Davis, J.M.; Himwich, W.A. Developmental changes in mouse brain: Weight, water content and free amino acids. *J. Neurochem.* **1968**, *15*, 917–923. [CrossRef]
18. Benítez-Diaz, P.; Miranda-Contreras, L.; Mendoza-Briceño, R.V.; Peña-Contreras, Z.; Palacios-Prü, E. Prenatal and postnatal contents of amino acid neurotransmitters in mouse parietal cortex. *Dev. Neurosci.* **2003**, *25*, 366–374. [CrossRef]
19. Sergeeva, O.A.; Chepkova, A.N.; Doreulee, N.; Eriksson, K.S.; Poelchen, W.; Mönnighoff, I.; Heller-Stilb, B.; Warskulat, U.; Häussinger, D.; Haas, H.L. Taurine-induced long-lasting enhancement of synaptic transmission in mice: Role of transporters. *J. Physiol.* **2003**, *550*, 911–919. [CrossRef]
20. Neuringer, M.; Sturman, J. Visual acuity loss in rhesus monkey infants fed a taurine-free human infant formula. *J. Neurosci. Res.* **1987**, *18*, 602–614. [CrossRef]
21. Neuringer, M.; Imaki, H.; Sturman, J.; Moretz, R.; Wisniewski, H. Abnormal visual acuity and retinal morphology in rhesus monkeys fed a taurine-free diet during the first three postnatal months. *Adv. Exp. Med. Biol.* **1987**, *217*, 125–134. [PubMed]
22. Kilb, W.; Fukuda, A. Taurine as an essential neuromodulator during perinatal cortical development. *Front. Cell. Neurosci.* **2017**, *11*, 328. [CrossRef] [PubMed]
23. Furukawa, T.; Yamada, J.; Akita, T.; Matsushima, Y.; Yanagawa, Y.; Fukuda, A. Roles of taurine-mediated tonic $GABA_A$ receptor activation in the radial migration of neurons in the fetal mouse cerebral cortex. *Front. Cell. Neurosci.* **2014**, *8*, 88. [CrossRef] [PubMed]
24. Tochitani, S.; Furukawa, T.; Bando, R.; Kondo, S.; Ito, T.; Matsushima, Y.; Kojima, T.; Matsuzaki, H.; Fukuda, A. $GABA_A$ receptors and maternally derived raurine regulate the temporal specification of progenitors of excitatory glutamatergic neurons in the mouse developing Cortex. *Cereb. Cortex* **2021**, *31*, 4554–4575. [CrossRef] [PubMed]
25. Inoue, K.; Furukawa, T.; Kumada, T.; Yamada, J.; Wang, T.; Inoue, R.; Fukuda, A. Taurine inhibits K^+-Cl^- cotransporter KCC2 to regulate embryonic Cl^- homeostasis via with-no-lysine (WNK) protein kinase signaling pathway. *J. Biol. Chem.* **2012**, *287*, 20839–20850. [CrossRef] [PubMed]
26. Warskulat, U.; Flögel, U.; Jacoby, C.; Hartwig, H.G.; Thewissen, M.; Merx, M.W.; Molojavyi, A.; Heller-Stilb, B.; Schrader, J.; Häussinger, D. Taurine transporter knockout depletes muscle taurine levels and results in severe skeletal muscle impairment but leaves cardiac function uncompromised. *FASEB J.* **2004**, *18*, 577–579. [CrossRef]
27. Ito, T.; Yoshikawa, N.; Schaffer, S.W.; Azuma, J. Tissue taurine depletion alters metabolic response to exercise and reduces running capacity in mice. *J. Amino Acids* **2014**, *2014*, 964680. [CrossRef]
28. Ito, T.; Nakanishi, Y.; Yamaji, N.; Murakami, S.; Schaffer, S.W. Induction of growth differentiation factor 15 in skeletal muscle of old taurine transporter knockout mouse. *Biol. Pharm. Bull.* **2018**, *41*, 435–439. [CrossRef]
29. Ito, T.; Yoshikawa, N.; Ito, H.; Schaffer, S.W. Impact of taurine depletion on glucose control and insulin secretion in mice. *J. Pharmacol. Sci.* **2015**, *129*, 59–64. [CrossRef]
30. Murakami, S. The physiological and pathophysiological roles of taurine in adipose tissue in relation to obesity. *Life Sci.* **2017**, *186*, 80–86. [CrossRef]
31. Guo, Y.Y.; Li, B.Y.; Peng, W.Q.; Guo, L.; Tang, Q.Q. Taurine-mediated browning of white adipose tissue is involved in its anti-obesity effect in mice. *J. Biol. Chem.* **2019**, *294*, 15014–15024. [CrossRef] [PubMed]
32. Jung, Y.; Choi, M. Relation of taurine intake during pregnancy and newborns' growth. *Adv. Exp. Med. Biol.* **2019**, *1155*, 283–292. [PubMed]
33. Filgueiras, G.B.; Carvalho-Netto, E.F.; Estanislau, C. Aversion in the elevated plus-maze: Role of visual and tactile cues. *Behav. Process.* **2014**, *107*, 106–111. [CrossRef] [PubMed]
34. Albrecht, J.; Schousboe, A. Taurine interaction with neurotransmitter receptors in the CNS: An update. *Neurochem. Res.* **2005**, *30*, 1615–1621. [CrossRef] [PubMed]
35. Flint, A.C.; Liu, X.; Kriegstein, A.R. Nonsynaptic glycine receptor activation during early neocortical development. *Neuron* **1998**, *20*, 43–53. [CrossRef]
36. Palackal, T.; Moretz, R.; Wisniewski, H.; Sturman, J.A. Abnormal visual cortex development in the kitten associated with maternal dietary taurine deprivation. *J. Neurosci. Res.* **1986**, *15*, 223–239. [CrossRef]
37. Sturman, J.A.; Moretz, R.C.; French, J.H.; Wisniewski, H.M. Taurine deficiency in the developing cat: Persistence of the cerebellar external granule cell layer. *J. Neurosci. Res.* **1985**, *13*, 405–416. [CrossRef]
38. Shivaraj, M.C.; Marcy, G.; Low, G.; Ryu, J.R.; Zhao, X.; Rosales, F.J.; Goh, E.L.K. Taurine induces proliferation of neural stem cells and synapse development in the developing mouse brain. *PLoS ONE* **2012**, *7*, e42935. [CrossRef]
39. Jia, F.; Yue, M.; Chandra, D.; Keramidas, A.; Goldstein, P.A.; Homanics, G.E.; Harrison, N.L. Taurine is a potent activator of extrasynaptic $GABA_A$ receptors in the thalamus. *J. Neurosci.* **2008**, *28*, 106–115. [CrossRef]

40. Sergeeva, O.A.; Fleischer, W.; Chepkova, A.N.; Warskulat, U.; Häussinger, D.; Siebler, M.; Haas, H.L. $GABA_A$-receptor modification in taurine transporter knockout mice causes striatal disinhibition. *J. Physiol.* **2007**, *585*, 539–548. [CrossRef]
41. Harding, N.J.; Davies, W.E. Cellular localisation of taurine in the organ of Corti. *Hear. Res.* **1993**, *65*, 211–215. [CrossRef]
42. Palkovits, M.; Elekes, I.; Láng, T.; Patthy, A. Taurine levels in discrete brain nuclei of rats. *J. Neurochem.* **1986**, *47*, 1333–1335. [CrossRef] [PubMed]
43. Lee, A.C.; Godfrey, D.A. Cochlear damage affects neurotransmitter chemistry in the central auditory system. *Front. Neurol.* **2014**, *5*, 227. [CrossRef] [PubMed]
44. Jiang, H.; Ding, D.; Müller, M.; Pfister, M.; Warskulat, U.; Häussinger, D.; Salvi, R. Central auditory nerve degeneration in knockout mice lacking the taurine transporter. In Proceedings of the 28th Annual Meeting of the Association for Research in Otolaryngology, New Orleans, LA, USA, 19–24 February 2005.
45. Xu, H.; Wang, W.; Tang, Z.Q.; Xu, T.L.; Chen, L. Taurine acts as a glycine receptor agonist in slices of rat inferior colliculus. *Hear. Res.* **2006**, *220*, 95–105. [CrossRef] [PubMed]
46. Song, N.Y.; Shi, H.B.; Li, C.Y.; Yin, S.K. Interaction between taurine and $GABA_A$/glycine receptors in neurons of the rat anteroventral cochlear nucleus. *Brain Res.* **2012**, *1472*, 1–10. [CrossRef] [PubMed]
47. Liu, H.; Du, T.; Li, C.; Yang, G. STAT3 phosphorylation in central leptin resistance. *Nutr. Metab.* **2021**, *18*, 39. [CrossRef] [PubMed]
48. Douros, J.D.; Baltzegar, D.A.; Reading, B.J.; Seale, A.P.; Lerner, D.T.; Gordon Grau, E.; Borski, R.J. Leptin stimulates cellular glycolysis through a STAT3 dependent mechanism in Tilapia. *Front. Endocrinol.* **2018**, *9*, 465. [CrossRef] [PubMed]
49. Yang, J.; Wu, G.; Feng, Y.; Lv, Q.; Lin, S.; Hu, J. Effects of taurine on male reproduction in rats of different ages. *J. Biomed. Sci.* **2010**, *17*, S9. [CrossRef]
50. Dawson, R., Jr.; Biasetti, M.; Messina, S.; Dominy, J. The cytoprotective role of taurine in exercise-induced muscle injury. *Amino Acids* **2002**, *22*, 309–324. [CrossRef]
51. Ito, T.; Okazaki, K.; Nakajima, D.; Shibata, D.; Murakami, S.; Schaffer, S. Mass spectrometry-based metabolomics to identify taurine-modified metabolites in heart. *Amino Acids* **2018**, *50*, 117–124. [CrossRef]
52. Ito, T.; Murakami, S.; Schaffer, S. Pathway analysis of a transcriptome and metabolite profile to elucidate a compensatory mechanism for taurine deficiency in the heart of taurine transporter knockout mice. *J* **2018**, *1*, 57–70. [CrossRef]
53. Guadagnin, E.; Mázala, D.; Chen, Y.W. STAT3 in skeletal muscle function and disorders. *Int. J. Mol. Sci.* **2018**, *19*, 2265. [CrossRef] [PubMed]
54. Wada, E.; Tanihata, J.; Iwamura, A.; Takeda, S.; Hayashi, Y.K.; Matsuda, R. Treatment with the anti-IL-6 receptor antibody attenuates muscular dystrophy via promoting skeletal muscle regeneration in dystrophin-/utrophin-deficient mice. *Skelet. Muscle* **2017**, *7*, 23. [CrossRef] [PubMed]
55. Nunes, A.M.; Wuebbles, R.D.; Sarathy, A.; Fontelonga, T.M.; Deries, M.; Burkin, D.J.; Thorsteinsdóttir, S. Impaired fetal muscle development and JAK-STAT activation mark disease onset and progression in a mouse model for merosin-deficient congenital muscular dystrophy. *Hum. Mol. Genet.* **2017**, *26*, 2018–2033. [CrossRef]

Article

Central Taurine Attenuates Hyperthermia and Isolation Stress Behaviors Augmented by Corticotropin-Releasing Factor with Modifying Brain Amino Acid Metabolism in Neonatal Chicks

Mohamed Z. Elhussiny [1,2], Phuong V. Tran [1], Yuriko Tsuru [1], Shogo Haraguchi [3], Elizabeth R. Gilbert [4], Mark A. Cline [4], Takashi Bungo [5], Mitsuhiro Furuse [1] and Vishwajit S. Chowdhury [1,6,*]

1 Laboratory of Regulation in Metabolism and Behavior, Graduate School of Bioresource and Bioenvironmental Science, Kyushu University, Fukuoka 819-0395, Japan; mohamedzakaria@vet.aswu.edu.eg (M.Z.E.); tran.viet.phuong.737@m.kyushu-u.ac.jp (P.V.T.); turu.yuriko.508@s.kyushu-u.ac.jp (Y.T.); furuse@brs.kyushu-u.ac.jp (M.F.)
2 Department of Animal & Poultry Behavior and Management, Faculty of Veterinary Medicine, Aswan University, Aswan 81528, Egypt
3 Department of Biochemistry, Showa University School of Medicine, Tokyo 142-8555, Japan; shogo@med.showa-u.ac.jp
4 School of Neuroscience, Virginia Polytechnic Institute and State University, Blacksburg, VA 24061-0306, USA; egilbert@vt.edu (E.R.G.); macline2@vt.edu (M.A.C.)
5 Department of Bioresource Science, Graduate School of Biosphere Science, Hiroshima University, Higashi-Hiroshima 739-8528, Japan; bungo@hiroshima-u.ac.jp
6 Division of Experimental Natural Science, Faculty of Arts and Science, Kyushu University, Fukuoka 819-0395, Japan
* Correspondence: vc-sur@artsci.kyushu-u.ac.jp

Citation: Elhussiny, M.Z.; Tran, P.V.; Tsuru, Y.; Haraguchi, S.; Gilbert, E.R.; Cline, M.A.; Bungo, T.; Furuse, M.; Chowdhury, V.S. Central Taurine Attenuates Hyperthermia and Isolation Stress Behaviors Augmented by Corticotropin-Releasing Factor with Modifying Brain Amino Acid Metabolism in Neonatal Chicks. *Metabolites* **2022**, *12*, 83. https://doi.org/10.3390/metabo12010083

Academic Editors: Teruo Miyazaki, Takashi Ito, Alessia Baseggio Conrado and Shigeru Murakami

Received: 21 December 2021
Accepted: 13 January 2022
Published: 16 January 2022

Abstract: The objective of this study was to determine the effects of centrally administered taurine on rectal temperature, behavioral responses and brain amino acid metabolism under isolation stress and the presence of co-injected corticotropin-releasing factor (CRF). Neonatal chicks were centrally injected with saline, 2.1 pmol of CRF, 2.5 μmol of taurine or both taurine and CRF. The results showed that CRF-induced hyperthermia was attenuated by co-injection with taurine. Taurine, alone or with CRF, significantly decreased the number of distress vocalizations and the time spent in active wakefulness, as well as increased the time spent in the sleeping posture, compared with the saline- and CRF-injected chicks. An amino acid chromatographic analysis revealed that diencephalic leucine, isoleucine, tyrosine, glutamate, asparagine, alanine, β-alanine, cystathionine and 3-methylhistidine were decreased in response to taurine alone or in combination with CRF. Central taurine, alone and when co-administered with CRF, decreased isoleucine, phenylalanine, tyrosine and cysteine, but increased glycine concentrations in the brainstem, compared with saline and CRF groups. The results collectively indicate that central taurine attenuated CRF-induced hyperthermia and stress behaviors in neonatal chicks, and the mechanism likely involves the repartitioning of amino acids to different metabolic pathways. In particular, brain leucine, isoleucine, cysteine, glutamate and glycine may be mobilized to cope with acute stressors.

Keywords: taurine; CRF; chicks; isolation stress; sedation; hypnosis

1. Introduction

Stress is a state of altered body homeostasis that is mediated through physiological changes and behavioral responses. While the stress response is crucial for animal adaptation to a novel environment, it is linked to anxiety-related disorders when chronically activated [1], and, in particular, social isolation stress has been more prevalent in recent years due to the COVID-19 pandemic [2,3]. Chicks serve as a model animal for anxiety, and when isolated from their social group, display predictable, quantifiable behaviors [4] that have been described [5,6]. For instance, an isolation stress paradigm is useful for screening anxiolytic drugs in chicks with recordings of spontaneous activity and distress vocalizations [5,7,8].

Corticotropin-releasing factor (CRF), a 41 amino acid polypeptide that originates from the paraventricular nucleus of the hypothalamus, modulates stress-related effects in the central nervous system (CNS), such as alteration in wakefulness and sleep [9,10]. This peptide has a variety of biological effects and is an indicator of hypothalamic–pituitary–adrenal (HPA) axis activity [11–13]. Studies of the CRF effects in animal models involve the administration of CRF or receptor antagonists into the brain, usually into the ventricular system (i.e., intracerebroventricular; ICV). CRF increased the magnitude of anxiety among socially separated/isolated chicks [14–16] and was involved in stress-induced hyperthermia (SIH) in rats [11,17,18]. In general, across species, CRF treatment induces strong anorectic and thermogenic effects [19], and the effects in chicks, including behavioral responses to social isolation stress, have been described [20,21]. Thus, CRF is a key mediator of stress responses in birds and mammals, with effects on food intake, body temperature, and metabolism likely serving to alter or restore whole-body homeostasis.

Core body temperature is a key indicator of the physiological state during the response to a stressor. SIH is stimulated in response to the activation of the HPA axis [22], and social isolation-SIH occurred in rats [23] and humans [24]. SIH was used to screen for anxiolytic activity in singly housed mice, in which rectal temperature was elevated relative to the group-housed mice [25]. Thus, body temperature is a useful marker for anxiolytic activity and drug screening.

The brain amino acid concentrations in neonatal chicks are altered under stressful conditions, including social isolation stress and fasting stress [8]. Hamasu et al. [8] reported that diencephalic alanine, arginine, asparagine, aspartic acid, phenylalanine, proline and serine were reduced in response to isolation-induced stress. Several amino acids have sedative and hypnotic effects in chicks, including L-proline [8], L-serine [26], L-ornithine [16] and L-aspartate [6]. Thus, it stands to reason that the attenuation of stress-related behaviors could be achieved through nutritional intervention with dietary amino acids. Taurine (2-aminoethane sulfonic acid) is found in high concentrations in mammalian tissues, including the brain [27]. Taurine plays a role in multiple physiological functions, including metabolic activity, antioxidation, osmoregulation, membrane stabilization and neurotransmission [27–29]. Taurine, an inhibitory neurotransmitter in the brain, has a positive allosteric modulatory effect on ligand-gated chloride channels in neurons, including the ionotropic γ-amino butyric acid receptor ($GABA_A$-R) and the glycine receptor, as well as inhibitory effects on other ligand- and voltage-gated cation channels, such as the *N*-methyl-*D*-aspartate receptor [30,31]. Our recent findings showed that the central injection of taurine caused hypothermia in a dose-dependent manner in chicks that was mediated through $GABA_A$-R [32]. In addition, taurine was reported to induce anxiolytic effects in mice and rats [33,34], and the consumption of taurine-supplemented diets was associated with antidepressant-like behaviors in forced swimming-tested mice [35]. However, there is no report on the use of taurine to minimize social isolation stress and SIH.

The hypothalamus is considered to be the central hub for thermoregulation [36]. Kataoka et al. [37] reported that the rostral modulatory raphe and the dorsomedial hypothalamus mediated SIH. However, the locus coeruleus (LC), a small brainstem nucleus, is considered the primary site for norepinephrine production in the brain, which is implicated in the etiology of anxiety or stress [38,39]. Furthermore, LC dendrites receive input from excitatory terminals containing CRF [40].

In this study, we assessed the effects of centrally administered taurine and CRF on rectal temperature and behavioral responses in a neonatal chick social isolation stress model. We also investigated the involvement of amino acids with a focus on the hypothalamus and brainstem, which contain nuclei that are known to be involved in regulating the stress response and SIH.

2. Results

2.1. Changes in Rectal Temperature after Social Isolation and ICV Injection of Taurine and CRF

The central injection of taurine significantly ($p < 0.0001$) reduced rectal temperatures, whereas CRF significantly ($p < 0.001$) increased temperatures, compared to the saline-injected controls (Figure 1A). There were interactions between taurine and time ($p < 0.0001$) and CRF and time ($p < 0.001$), which demonstrate the hypo- and hyperthermia induced by taurine and CRF injections, respectively. When taurine and CRF were co-administered, there was no effect on temperature.

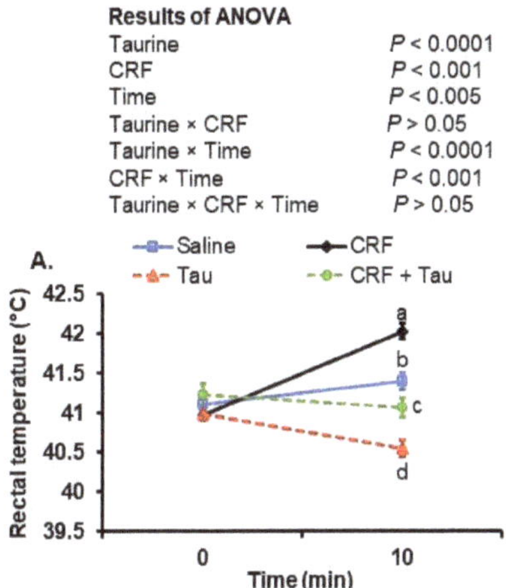

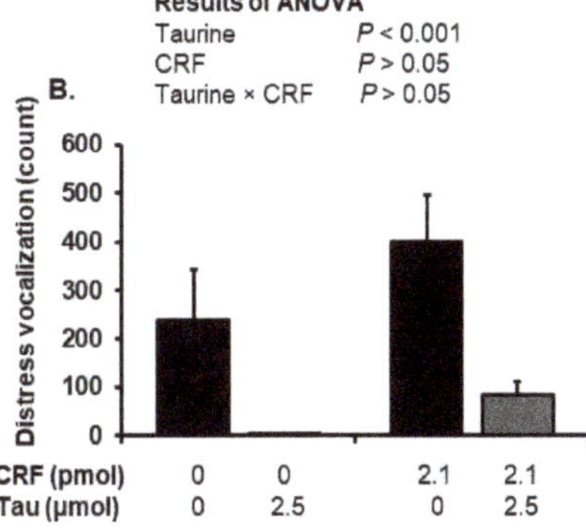

Figure 1. Rectal temperature (**A**) and distress vocalization (**B**) of chicks following the ICV injection of saline (control), taurine (2.5 μmol), CRF (2.1 pmol) or CRF plus taurine under social isolation stress. Values are expressed as mean ± SEM from chick groups (8–10). Means with different superscripts indicate statistically significant differences ($p < 0.05$). Tau, taurine; CRF, corticotropin-releasing factor.

2.2. Changes in Distress Vocalizations after Social Isolation Stress and Injection of Taurine and CRF

The injection of taurine with or without a co-injection of CRF was associated with a significant ($p < 0.001$) decrease in the number of distress vocalizations compared to the saline- and CRF-injected chicks (Figure 1B).

2.3. Behavioral Alterations after Social Isolation Stress and Injection of Taurine and CRF

The injection of taurine with or without CRF significantly ($p < 0.05$) decreased the time spent in active wakefulness compared to the saline- and CRF-injected chicks (Figure 2A). Compared to the CRF- and saline-injected chick groups, taurine significantly ($p < 0.0001$) increased the time spent in the sleeping posture when injected alone or with CRF. However, CRF significantly ($p < 0.05$) decreased the time spent in the sleeping posture, regardless of a co-injection with taurine, compared to the saline- and taurine-injected chick groups (Figure 2B). No significant changes were detected in the time spent standing or sitting motionless with the eyes opened or the time spent standing motionless with the eyes closed (Figure 2C,D).

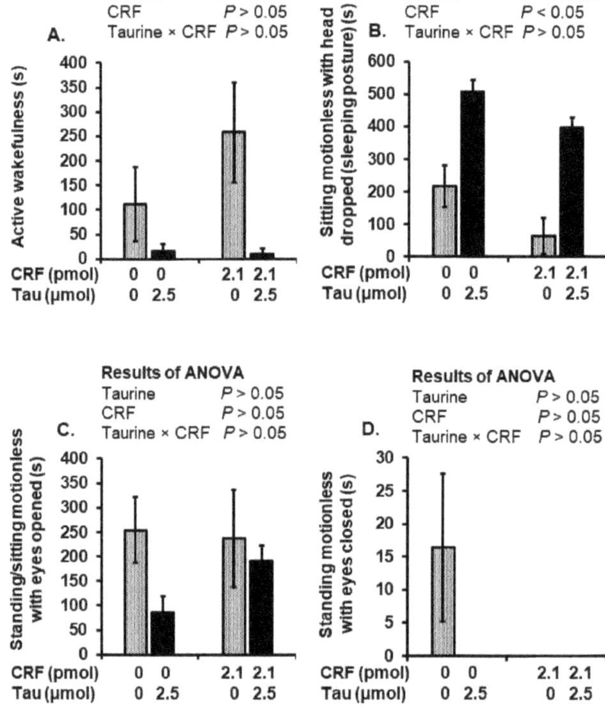

Figure 2. Active wakefulness (**A**), sitting motionless with the head dropped (sleeping posture) (**B**), standing/sitting motionless with eyes open (**C**) and standing motionless with the eyes closed (**D**) of the chicks following an ICV injection of saline (control), taurine (2.5 μmol), CRF (2.1 pmol) or CRF plus taurine under social isolation stress. Values are expressed as mean ± SEM from the chick groups (9–10). Means with different superscripts indicate statistically significant differences ($p < 0.05$). Tau, taurine; CRF, corticotropin-releasing factor.

2.4. Brain Amino Acid Changes after Social Isolation Stress and Injection of Taurine and CRF

Diencephalic leucine, isoleucine, tyrosine, glutamate, asparagine, alanine, β-alanine, cystathionine and 3-methylhistidine concentrations significantly ($p < 0.05$) decreased in taurine- and taurine–CRF-co-injected chicks compared with saline- and CRF-injected chicks (Table 1). Diencephalic taurine concentrations significantly ($p < 0.0001$) increased in taurine–CRF-co-injected chicks compared with saline- and CRF-injected chicks. The other diencephalic amino acid concentrations are shown in Supplementary Table S1.

Table 1. Effect of the intracerebroventricular injection of taurine or CRF on the amino acid concentrations in the diencephalon of chicks exposed to social isolation stress.

Free Amino Acids	Saline	Taurine	CRF	Taurine + CRF	*p* Value Taurine	*p* Value CRF	*p* Value Taurine × CRF
Essential amino acids							
Leucine	221 ± 6	198 ± 7	212 ± 7	208 ± 2	$p < 0.05$	NS	NS
Isoleucine	115 ± 5	97 ± 4	113 ± 6	100 ± 4	$p < 0.005$	NS	NS
Nonessential amino acids							
Taurine	4183 ± 57	5863 ± 162	4232 ± 101	5276 ± 263	$p < 0.0001$	NS	NS
Tyrosine	105 ± 5	101 ± 4	121 ± 7	102 ± 5	$p < 0.05$	NS	NS
Glutamic acid	6650 ± 69	6308 ± 68	6592 ± 144	6246 ± 119	$p < 0.005$	NS	NS
Asparagine	374 ± 7	349 ± 10	383 ± 7	360 ± 11	$p < 0.05$	NS	NS
Alanine	893 ± 15	860 ± 21	897 ± 29	838 ± 9	$p < 0.05$	NS	NS
β-Alanine	319 ± 16	298 ± 12	318 ± 8	286 ± 11	$p < 0.05$	NS	NS
Cystathionine	16 ± 1	15 ± 1	16 ± 1	14 ± 1	$p < 0.05$	NS	NS
3-Methylhistidine	304 ± 32	277 ± 25	323 ± 21	238 ± 19	$p < 0.05$	NS	NS

The number of chicks used in each group was 7–9. Values are means ± SEM in pmol/mg wet tissue. NS, not significant; CRF, corticotropin-releasing factor.

The brainstem concentrations of isoleucine, phenylalanine, tyrosine and cysteine significantly ($p < 0.05$) decreased in the taurine–CRF-co-injected chicks compared with the saline- and CRF-injected chicks (Table 2). The injection of taurine and co-injection with CRF significantly ($p < 0.05$) increased glycine concentrations. Taurine concentrations were significantly ($p < 0.0005$) elevated after the taurine and CRF co-injection. The interactions indicate that taurine and tyrosine were differentially affected after the injection of CRF or taurine. Other brainstem amino acid concentrations are shown in Supplementary Table S2.

Table 2. Effect of the intracerebroventricular injection of taurine or CRF on amino acid concentrations in the brainstem of chicks exposed to social isolation stress.

Free Amino Acids	Saline	Taurine	CRF	Taurine + CRF	*p* Value Taurine	*p* Value CRF	*p* Value Taurine × CRF
Essential amino acids							
Isoleucine	93 ± 4	89 ± 2	95 ± 2	84 ± 4	$p < 0.05$	NS	NS
Phenylalanine	88 ± 4	84 ± 6	95 ± 6	75 ± 2	$p < 0.05$	NS	NS
Glycine	2814 ± 66	2955 ± 62	2953 ± 63	3067 ± 44	$p < 0.05$	$p < 0.05$	NS
Nonessential amino acids							
Taurine	2686 ± 66 ^{b}	3903 ± 253 ^{a}	2982 ± 127 ^{b}	3168 ± 90 ^{b}	$p < 0.0005$	NS	$p < 0.005$
Tyrosine	86 ± 5 ^{ab}	88 ± 4 ^{ab}	107 ± 8 ^{a}	81 ± 4 ^{b}	$p < 0.05$	NS	$p < 0.05$
Cysteine	100 ± 5	88 ± 4	93 ± 5	86 ± 5	$p < 0.05$	NS	NS

The number of chicks used in each group was 7–9. Values are means ± SEM in pmol/mg wet tissue. Different superscripts in the same row indicate significant differences ($p < 0.05$) between saline, taurine, CRF and CRF + taurine groups. NS, not significant; CRF, corticotropin-releasing factor.

3. Discussion

In this study, we investigated the effects of social isolation stress, central taurine and CRF on rectal temperature, behavioral responses and brain amino acid concentrations. The ICV injection of CRF caused hyperthermia, which was attenuated by a co-injection with taurine. CRF induces hyperthermia through the activation of the HPA axis [41] and the dose-dependently increased O_2 consumption, CO_2 production and heat production in the chicks [42]. On the other hand, the central injection of taurine induced dose-dependent hypothermia in chicks [32] and in rabbits [43]. It could be speculated that taurine-induced hypothermia and attenuated CRF magnified SIH through a reduction in heat production. Further study is needed to elucidate how taurine antagonizes CRF-induced hyperthermia, although we did observe changes in some behaviors and brain amino acids that may help explain such effects. For instance, central taurine decreased the number of distress vocalizations, decreased the time spent in active wakefulness and increased the time spent in the sleeping posture. Distress vocalization is considered to be an important feature of the stress response during isolation [5,44].

Glutamate, a non-essential amino acid, is locally synthetized from glutamine or Krebs cycle intermediates and acts as an excitatory neurotransmitter in the CNS [45]. Glutamate in the brain is made locally due to low permeability at the blood–brain barrier (BBB) [46]; a very small amount of glutamate can pass through the BBB under high blood glutamate concentrations [47]. In addition, leucine and isoleucine, branched-chain amino acids (BCAAs), are important sources of nitrogen for glutamate synthesis [48–51]. Leucine increased glutamate dehydrogenase and glutamate synthesis from different nitrogen sources in the liver of pigs [52]. Alanine donates its amino group to α-ketoglutarate by alanine transaminase to form glutamate [53]. Thus, decreased leucine, isoleucine and alanine concentrations may reflect their contribution to the synthesis of glutamate, which was decreased in the present study in response to the taurine injection. The reason for diminished glutamate could be connected to the synthesis of glutathione (GSH), a glutamate-containing tripeptide (γ-glutamyl-cysteine-glycine). Glutamate and cysteine are used as substrates by γ-Glu-Cys synthetase to produce γ-glutamyl-cysteine dipeptide (γ-Glu-Cys), which is then combined with glycine to produce GSH [54,55]. Central GSH-induced sedation and hypnosis in neonatal chicks [56]. Furthermore, Glu-Cys and Glu-Gly dipeptides [55] and glutamate [57] decreased distress vocalizations and the time spent in active wakefulness in a dose-dependent manner in neonatal chicks. Cystathionine is synthesized from homocysteine and serine and then metabolized to cysteine by cystathionine γ-lyase to form GSH [58]. Decreased brain glutamate, cystathionine and cysteine in taurine-injected chicks suggest that these amino acids may be utilized for GSH synthesis to induce sedation and hypnosis. ICV glycine produced sedative effects in neonatal chicks that were mediated by the glycine receptor [26]. Although it is difficult to hypothesize why glycine concentrations were increased in the current study, taurine may stimulate the glycine metabolism to induce sedation and hypnosis.

Tyrosine is the precursor to catecholamine biosynthesis in the brain. It is converted into L-DOPA via tyrosine hydroxylase and then into dopamine and norepinephrine through two sequential enzymatic reactions pre-synaptically [59,60]. Our recent study showed that a higher dose (5 μmol) of ICV taurine stimulated catecholamine biosynthesis in the regulation of body temperature in neonatal chicks (unpublished data). In the present study, taurine-injected chicks, irrespective of CRF injection, had less diencephalic and brainstem tyrosine. Thus, taurine may stimulate the tyrosine metabolism to produce catecholamines, which may be involved in the regulation of body temperature and stress response in chicks.

Taurine and β-alanine were reported to be antagonistic to each other at the BBB in terms of transport because they are both in the β-amino acid category [61]. The chronic supplementation of β-alanine decreased brain taurine concentrations [62]. Increased brain taurine could thereby lead to a decrease in β-alanine concentrations in the brain.

In conclusion, the ICV injection of taurine attenuates CRF-induced hyperthermia and isolation stress behaviors in neonatal chicks. Brain leucine, isoleucine, glutamate, cysteine, cystathionine and glycine may be utilized in GSH synthesis to regulate the stress response through inducing sedative and hypnotic effects. These results suggest that taurine may serve as a novel isolation stress-relieving agent. Further study is needed to determine whether such effects are achievable through a dietary or other peripheral routes of administration.

4. Materials and Methods

The experimental procedures were conducted in accordance with the guidelines for animal experiments of the Faculty of Agriculture and the Graduate Course of Kyushu University and complied with Law No. 105 and Notification No. 6 of the Japanese government. The experimental protocol was approved by the animal experiment committee in Kyushu University (authorization no. A20-282-2).

4.1. Animals

Fertilized eggs (Julia layer strain, *Gallus gallus domesticus*) from a nearby hatchery (Tsuboi hatchery, Kumamoto, Japan) were incubated at 37.6 °C with 60% relative humidity in an incubator (Rcom Maru Deluxe MAX 380, Autoelex Co., Ltd., Gimhae-si, Korea) with auto-turning every hour until day 18. On embryonic day 7, the eggs were candled, and undeveloped and dead embryos were discarded. On embryonic day 19, the eggs were transferred to hatching trays. After hatching, day-old layer chicks were reared in groups with 20 birds per metal cage (floor space: 50 cm × 35 cm; height: 33 cm) under thermoneutral temperature (CT; 30 ± 1 °C) and continuous light. Feed (Adjust diet; Toyohashi Feed Co., Ltd., Aichi, Japan; metabolizable energy: >12.55 MJ/kg, protein > 23%) and water were provided *ad libitum*. On the second day after hatching, chicks were feather-sexed and male chicks were separated and used for the experiments. At four days old, chicks were assigned to treatment groups based on body weight.

4.2. Injection Procedure

The taurine was purchased from Wako Pure Industries, Ltd. (Osaka, Japan). The rat CRF was purchased from Peptide Institute, Inc. (Osaka, Japan). The taurine and CRF were dissolved in 0.85% saline containing 0.1% Evans blue. Evans blue solution was injected into the control group chicks, as described [32,63]. The drugs were incubated on ice during the experiments. The ICV injection was performed at five days of age, using a microsyringe, as described by Davis et al. [64]. Briefly, the head of the chick was introduced into an acrylic device that provides an injection site guide, and the solutions were injected into the left lateral ventricle, and the syringe was held in place for 10 s to prevent overflow. The ICV injection procedure is not stressful for the chicks [65].

4.3. Experimental Design

On the day of the experiments, the chicks (five days old; n = 10) were injected with 10 μL of saline, CRF (2.1 pmol [6,16]), taurine (2.5 μmol [32]), or CRF plus taurine (2.1 pmol + 2.5 μmol). The chicks were then individually housed in an acrylic glass chamber (40 cm × 30 cm × 20 cm) at a temperature of 30 ± 1 °C, and water and diet were withheld. The vocalizations and behaviors were audio and video recorded from three different directions for 10 min using digital cameras (JVC, Everio, Japan). Behaviors were classified into four categories, including active wakefulness, standing/sitting motionless with the eyes open, standing motionless with the eyes closed and sitting motionless with the head dropped (sleeping posture) [66]. The vocalizations were counted using a digital hand tally counter, and the behaviors were analyzed by a researcher blind to treatment. A digital thermometer with a precision of ±0.1 °C (Thermalert TH-5, Physitemp Instruments Inc., Clifton, NJ, USA) was used to measure the rectal temperature at 0 and 10 min post-injection by inserting the thermistor probe into the rectum via the cloaca to a depth of approximately 2 cm, as we reported [32,67]. The chicks were anesthetized 10 min post-ICV injection with isoflurane (Mylan Pharmaceutical Co., Ltd., Tokyo, Japan) before the collection of blood samples from the jugular vein. The chick brains were dissected following euthanasia, and the diencephalons (thalamus and hypothalamus) and brainstems were collected as described in the chick brain atlas [68]. The brain samples were snap-frozen and kept at −80°C until amino acid analysis. After opening the skulls for the brain dissections, the presence of dye in the ventricular system was verified to confirm the correct site of injection. Data for chicks lacking dye were excluded from the data analysis.

4.4. Brain Amino Acid Analysis

The free amino acids were analyzed in the diencephalon and brainstem using high-performance liquid chromatography (HPLC), according to the previous methods [69] with slight modifications [70]. The samples were briefly homogenized in ice-cold 0.2 M perchloric acid containing 0.01 mM ethylenediaminetetraacetic acid disodium salt (EDTA.2Na) and left on ice. After 30 min, the homogenates were centrifuged at 20,000× g for 15 min at

4 °C. After centrifugation, the supernatants were collected and filtered through 0.2 μm hydrophilic polytetrafluoroethylene filters (Millipore, Bedford, MA, USA). L-amino acid solutions (type ANII, type B, L-asparagine, L-glutamine, and L-tryptophan; Wako, Osaka, Japan) were used to prepare 200 pmol/μL of the standard solution. The standard solutions (10 μL) and brain tissue filtrate (20 μL) were adjusted to a pH of 7 with 1 M sodium hydroxide and dried under reduced pressure (Centrifugal Evaporator, CVE-3000, EYELA, Tokyo, Japan). Then, 10 μL of 1 mol/L sodium acetate-methanol-triethylamine (2:2:1) were used to dissolve dried residues, and the samples were re-dried under reduced pressure, followed by adding 20 μL of methanol-distilled water-triethylamine-phenylisothiocyanate (7:1:1:1). The samples were incubated for 20 min at room temperature to form phenyl thiocarbamoyl derivatives. The standard and samples were subjected to drying, again under reduced pressure, before dissolving in 200 μL of Pico-Tag Diluent (Waters, Milford, CT, USA), and a 0.20 μm filter (Millipore, Bedford, MA, USA) was used to obtain the filtrate. A Waters HPLC system (Pico-Tag free amino acid analysis column (3.9 mm × 300 mm), an Alliance e2695 Separations Module and a Waters 2487 dual-wavelength UV detector and Empower software) was used to apply the solution containing the derivatives. A gradient linear elution program (0%, 3%, 6%, 9%, 40% and 100%) was applied using mobile phase A and B at a flow rate of 1 mL/min at 46 °C. Mobile phase A consisted of 70 mmol/L of sodium acetate trihydrate and acetonitrile at a ratio of 975:25. The sodium acetate solution was adjusted to a pH of 6.45 by adding 10% acetic acid and was then filtered through a 0.45 μm MCE membrane (MF-Millipore, Merck Millipore Ltd., Cork, Ireland). Mobile phase B consisted of water, acetonitrile and methanol at a ratio of 40:45:15. The UV wavelength was set at 254 nm to determine the concentrations of amino acids. The concentrations of amino acids in the brain samples were expressed as pmol/mg, wet tissue. The system did not distinguish between L- and D-isomers; thus, only names of amino acids are provided.

4.5. Statistical Analyses

The rectal temperature results were analyzed using a repeated measure three-way analysis of variance (ANOVA), where the main effects were taurine, CRF and time, followed by the Tukey–Kramer test as a post hoc analysis. A two-way ANOVA was carried out for distress vocalizations, behavioral results and amino acids, where the main effects were taurine and CRF, followed by the Tukey–Kramer test as a post hoc analysis when a significant interaction was detected. The differences were considered significant at $p < 0.05$. The data are presented as means ± standard error of the mean (SEM). The Thompson rejection test was applied to eliminate the experimental data that contained outliers ($p < 0.01$) [71]. The statistical analysis was performed using StatView version 5.0 (SAS Institute Inc., Cary, NC, USA).

Supplementary Materials: The following are available online at https://www.mdpi.com/article/10.3390/metabo12010083/s1, Table S1: Effect of an intracerebroventricular injection of taurine or CRF on amino acid concentrations in the diencephalon of chicks exposed to social isolation stress; Table S2: Effect of an intracerebroventricular injection of taurine or CRF on amino acid concentrations in the brainstem of chicks exposed to social isolation stress.

Author Contributions: Conceptualization, V.S.C., M.F. and M.Z.E.; methodology, M.Z.E., P.V.T., Y.T. and V.S.C.; software, M.Z.E. and V.S.C.; validation, V.S.C., M.F. and M.Z.E.; formal analysis, M.Z.E. and V.S.C.; investigation, M.Z.E. and V.S.C.; resources, M.F. and V.S.C.; data curation, M.Z.E. and V.S.C.; writing—original draft preparation, M.Z.E. and V.S.C.; writing—review and editing, M.Z.E., S.H., E.R.G., M.A.C., T.B., M.F. and V.S.C.; visualization, M.Z.E. and V.S.C.; supervision, V.S.C.; project administration, V.S.C.; funding acquisition, V.S.C. All authors have read and agreed to the published version of the manuscript.

Funding: M.Z.E. was funded by a full scholarship from the Ministry of Higher Education of the Arab Republic of Egypt. This research was funded by JSPS KAKENHI, Grant Number JP19H03110, granted to V.S.C.

Institutional Review Board Statement: The animal study protocol was approved by ethics committee in Kyushu University (authorization no. A20-282-2).

Informed Consent Statement: Not applicable.

Data Availability Statement: Data are contained within the article.

Conflicts of Interest: The authors declare that they have no conflict of interest.

References

1. Golla, A.; Østby, H.; Kermen, F. Chronic unpredictable stress induces anxiety-like behaviors in young zebrafish. *Sci. Rep.* **2020**, *10*, 10339. [CrossRef]
2. Hwang, T.J.; Rabheru, K.; Peisah, C.; Reichman, W.; Ikeda, M. Loneliness, and social isolation during the COVID-19 pandemic. *Int. Psychogeriatr.* **2020**, *32*, 1217–1220. [CrossRef]
3. Sokolowska, J.; Ayton, P.; Brandstätter, E. Editorial: Coronavirus Disease (COVID-19): Psychological Reactions to the Pandemic. *Front. Psychol.* **2021**, *12*, 745941. [CrossRef] [PubMed]
4. Furuse, M. Screening of central functions of amino acids and their metabolites for sedative and hypnotic effects using chick models. *Eur. J. Pharmacol.* **2015**, *762*, 382–393. [CrossRef] [PubMed]
5. Feltenstein, M.W.; Lambdin, L.C.; Ganzera, M.; Ranjith, H.; Dharmaratne, W.; Nanayakkara, N.P.; Khan, I.A.; Sufka, K.J. Anxiolytic properties of Piper methysticum extract samples and fractions in the chick social-separation-stress procedure. *Phytother. Res.* **2003**, *17*, 210–216. [CrossRef]
6. Erwan, E.; Tomonaga, S.; Yoshida, J.; Nagasawa, M.; Ogino, Y.; Denbow, D.M.; Furuse, M. Central administration of L- and D-aspartate attenuates stress behaviors by social isolation and CRF in neonatal chicks. *Amino Acids* **2012**, *43*, 1969–1976. [CrossRef]
7. Feltenstein, M.W.; Warnick, J.E.; Guth, A.N.; Sufka, K.J. The chick separation stress paradigm: A validation study. *Pharmacol. Biochem. Behav.* **2004**, *77*, 221–226. [CrossRef] [PubMed]
8. Hamasu, K.; Haraguchi, T.; Kabuki, Y.; Adachi, N.; Tomonaga, S.; Sato, H.; Denbow, D.M.; Furuse, M. L-Proline is a sedative regulator of acute stress in the brain of neonatal chicks. *Amino Acids* **2009**, *37*, 377–382. [CrossRef] [PubMed]
9. Koob, G.F. Corticotropin-releasing factor, norepinephrine, and stress. *Biol. Psychiatry* **1999**, *46*, 1167–1180. [CrossRef]
10. Wellman, L.L.; Yang, L.; Sanford, L.D. Effects of corticotropin releasing factor (CRF) on sleep and temperature following predictable controllable and uncontrollable stress in mice. *Front. Neurosci.* **2015**, *9*, 258. [CrossRef]
11. Owens, M.J.; Nemeroff, C.B. Physiology and pharmacology of corticotropin-releasing factor. *Pharmacol. Rev.* **1991**, *43*, 425–473. [PubMed]
12. Furuse, M.; Matsumoto, M.; Saito, N.; Sugahara, K.; Hasegawa, S. The central corticotropin-releasing factor and glucagon-like peptide-1 in food intake of the neonatal chick. *Eur. J. Pharmacol.* **1997**, *339*, 211–214. [CrossRef]
13. Arborelius, L.; Owens, M.J.; Plotsky, P.M.; Nemeroff, C.B. The role of corticotropin-releasing factor in depression and anxiety disorders. *J. Endocrinol.* **1999**, *160*, 1–12. [CrossRef]
14. Zhang, R.; Tachibana, T.; Takagi, T.; Koutoku, T.; Denbow, D.M.; Furuse, M. Centrally administered norepinephrine modifies the behavior induced by corticotrophin-releasing factor in neonatal chicks. *J. Neurosci. Res.* **2003**, *74*, 630–636. [CrossRef]
15. Zhang, R.; Tachibana, T.; Takagi, T.; Koutoku, T.; Denbow, D.M.; Furuse, M. Serotonin modifies corticotrophin-releasing factor-induced behaviors of chicks. *Behav. Brain Res.* **2004**, *151*, 47–52. [CrossRef]
16. Kurata, K.; Shigemi, K.; Tomonaga, S.; Aoki, M.; Morishita, K.; Denbow, D.M.; Furuse, M. L-Ornithine attenuates corticotropin-releasing factor-induced stress responses acting at $GABA_A$ receptors in neonatal chicks. *Neuroscience* **2011**, *172*, 226–231. [CrossRef]
17. Nakamori, T.; Morimoto, A.; Murakami, N. Effect of a central CRF antagonist on cardiovascular and thermoregulatory responses induced by stress or IL-1 beta. *Am. J. Physiol.* **1993**, *265*, 834–839. [CrossRef] [PubMed]
18. Fabricio, A.S.; Rae, G.A.; Zampronio, A.R.; D'Orléans-Juste, P.; Souza, G.E. Central endothelin ET(B) receptors mediate IL-1-dependent fever induced by preformed pyrogenic factor and corticotropin-releasing factor in the rat. *Am. J. Physiol. Regul. Integr. Comp. Physiol.* **2006**, *290*, R164–R171. [CrossRef]
19. Richard, D.; Lin, Q.; Timofeeva, E. The corticotropin-releasing factor family of peptides and CRF receptors: Their roles in the regulation of energy balance. *Eur. J. Pharmacol.* **2002**, *440*, 189–197. [CrossRef]
20. Ohgushi, A.; Bungo, T.; Shimojo, M.; Masuda, Y.; Denbow, D.M.; Furuse, M. Relationships between feeding and locomotion behaviors after central administration of CRF in chicks. *Physiol. Behav.* **2001**, *72*, 287–289. [CrossRef]
21. Zhang, R.; Nakanishi, T.; Ohgushi, A.; Ando, R.; Yoshimatsu, T.; Denbow, D.M.; Furuse, M. Interaction of corticotropin-releasing factor and glucagon-like peptide-1 on behaviors in chicks. *Eur. J. Pharmacol.* **2001**, *430*, 73–78. [CrossRef]
22. Blache, D.; Terlouw, C.; Maloney, S.K. Physiology. In *Animal Welfare*, 3rd ed.; Appleby, M.C., Olsson, I.A.S., Galindo, F., Eds.; CABI: Wallingford, UK, 2017; pp. 181–212.
23. Kataoka, N.; Shima, Y.; Nakajima, K.; Nakamura, K. A central master driver of psychosocial stress responses in the rat. *Science* **2020**, *367*, 1105–1112. [CrossRef]
24. Vinkers, C.H.; Penning, R.; Hellhammer, J.; Verster, J.C.; Klaessens, J.H.; Olivier, B.; Kalkman, C.J. The effect of stress on core and peripheral body temperature in humans. *Stress* **2013**, *16*, 520–530. [CrossRef]

25. Van der Heyden, J.A.; Zethof, T.J.; Olivier, B. Stress-induced hyperthermia in singly housed mice. *Physiol. Behav.* **1997**, *62*, 463–470. [CrossRef]
26. Shigemi, K.; Tsuneyoshi, Y.; Hamasu, K.; Han, L.; Hayamizu, K.; Denbow, D.M.; Furuse, M. l-Serine induces sedative and hypnotic effects acting at GABA(A) receptors in neonatal chicks. *Eur. J. Pharmacol.* **2008**, *599*, 86–90. [CrossRef]
27. Wu, J.Y.; Prentice, H. Role of taurine in the central nervous system. *J. Biomed. Sci.* **2010**, *17*, S1. [CrossRef] [PubMed]
28. Schaffer, S.W.; Jong, C.J.; Ramila, K.C.; Azuma, J. Physiological roles of taurine in heart and muscle. *J. Biomed. Sci.* **2010**, *17*, S2. [CrossRef] [PubMed]
29. Zhang, Z.; Zhao, L.; Zhou, Y.; Lu, X.; Wang, Z.; Wang, J.; Li, W. Taurine ameliorated homocysteine induced H9C2 cardiomyocyte apoptosis by modulating endoplasmic reticulum stress. *Apoptosis* **2017**, *22*, 647–661. [CrossRef]
30. Olive, M.F. Interactions between taurine and ethanol in the central nervous system. *Amino Acids* **2002**, *23*, 345–357. [CrossRef]
31. Ochoa-de la Paz, L.; Zenteno, E.; Gulias-Cañizo, R.; Quiroz-Mercado, H. Taurine and GABA neurotransmitter receptors, a relationship with therapeutic potential. *Expert Rev. Neurother.* **2019**, *19*, 289–291. [CrossRef] [PubMed]
32. Elhussiny, M.Z.; Tran, P.V.; Pham, C.V.; Nguyen, L.T.N.; Haraguchi, S.; Gilbert, E.R.; Cline, M.A.; Bungo, T.; Furuse, M.; Chowdhury, V.S. Central $GABA_A$ receptor mediates taurine-induced hypothermia and possibly reduces food intake in thermo-neutral chicks and regulates plasma metabolites in heat-exposed chicks. *J. Therm. Biol.* **2021**, *98*, 102905. [CrossRef]
33. Chen, S.W.; Kong, W.X.; Zhang, Y.J.; Li, Y.L.; Mi, X.J.; Mu, X.S. Possible anxiolytic effects of taurine in the mouse elevated plus-maze. *Life Sci.* **2004**, *75*, 1503–1511. [CrossRef] [PubMed]
34. Kong, W.X.; Chen, S.W.; Li, Y.L.; Zhang, Y.J.; Wang, R.; Min, L.; Mi, X. Effects of taurine on rat behaviors in three anxiety models. *Pharmacol. Biochem. Behav.* **2006**, *83*, 271–276. [CrossRef]
35. Murakami, T.; Furuse, M. The impact of taurine- and beta-alanine-supplemented diets on behavioral and neurochemical parameters in mice: Antidepressant versus anxiolytic-like effects. *Amino Acids* **2010**, *39*, 427–434. [CrossRef]
36. Romanovsky, A.A. Thermoregulation: Some concepts have changed. Functional architecture of the thermoregulatory system. *Am. J. Physiol. Regul. Integr. Comp. Physiol.* **2007**, *292*, 37–46. [CrossRef]
37. Kataoka, N.; Hioki, H.; Kaneko, T.; Nakamura, K. Psychological stress activates a dorsomedial hypothalamus-medullary raphe circuit driving brown adipose tissue thermogenesis and hyperthermia. *Cell Metab.* **2014**, *20*, 346–358. [CrossRef] [PubMed]
38. Reyes, B.A.S.; Zitnik, G.; Foster, C.; van Bockstaele, E.J.; Valentino, R.J. Social stress engages neurochemically-distinct afferents to the rat locus coeruleus depending on coping strategy. *eNeuro* **2015**, *2*, 4215. [CrossRef] [PubMed]
39. Breton-Provencher, V.; Drummond, G.T.; Sur, M. Locus coeruleus norepinephrine in learned behavior: Anatomical modularity and spatiotemporal integration in targets. *Front. Neural Circuits* **2021**, *15*, 7. [CrossRef]
40. Tjoumakaris, S.I.; Rudoy, C.; Peoples, J.; Valentino, R.J.; van Bockstaele, E.J. Cellular interactions between axon terminals containing endogenous opioid peptides or corticotropin-releasing factor in the rat locus coeruleus and surrounding dorsal pontine tegmentum. *J. Comp. Neurol.* **2003**, *466*, 445–456. [CrossRef] [PubMed]
41. Tachibana, T.; Saito, E.S.; Saito, S.; Tomonaga, S.; Denbow, D.M.; Furuse, M. Comparison of brain arginine-vasotocin and corticotrophin-releasing factor for physiological responses in chicks. *Neurosci. Lett.* **2004**, *360*, 165–169. [CrossRef]
42. Tachibana, T.; Takahashi, H.; Oikawa, D.; Denbow, D.M.; Furuse, M. Thyrotropin-releasing hormone increased heat production without the involvement of corticotropin-releasing factor in neonatal chicks. *Pharmacol. Biochem. Behav.* **2006**, *83*, 528–532. [CrossRef]
43. Harris, W.S.; Lipton, J.M. Intracerebroventricular taurine in rabbits: Effects of normal body temperature, endotoxin fever and hyperthermia produced by PGE1 and amphetamine. *J. Physiol.* **1977**, *266*, 397–410. [CrossRef]
44. Yoshida, J.; Tomonaga, S.; Ogino, Y.; Nagasawa, M.; Kurata, K.; Furuse, M. Intracerebroventricular injection of kynurenic acid attenuates corticotrophin-releasing hormone-augmented stress responses in neonatal chicks. *Neuroscience* **2012**, *220*, 142–148. [CrossRef] [PubMed]
45. Smith, Q.R. Glutamate, and glutamine in the brain transport of glutamate and other amino acids at the blood-brain barrier. *J. Nutr.* **2000**, *130*, 1016. [CrossRef]
46. Bai, W.; Zhou, Y.G. Homeostasis of the intraparenchymal-blood glutamate concentration gradient: Maintenance, imbalance, and regulation. *Front. Mol. Neurosci.* **2017**, *10*, 400. [CrossRef]
47. Cederberg, H.H.; Uhd, N.C.; Brodin, B. Glutamate efflux at the blood–brain barrier: Cellular mechanisms and potential clinical relevance. *Arch. Med. Res.* **2014**, *45*, 639–645. [CrossRef] [PubMed]
48. Yudkoff, M.; Daikhin, Y.; Grunstein, L.; Nissim, I.; Stern, J.; Pleasure, D.; Nissim, I. Astrocyte leucine metabolism: Significance of branched-chain amino acid transamination. *J. Neurochem.* **1996**, *66*, 378–385. [CrossRef]
49. Bixel, M.G.; Hutson, S.M.; Hamprecht, B. Cellular distribution of branched-chain amino acid aminotransferase isoenzymes among rat brain glial cells in culture. *J. Histochem. Cytochem.* **1997**, *45*, 685–694. [CrossRef] [PubMed]
50. Hutson, S.M.; Lieth, E.; LaNoue, K.F. Function of leucine in excitatory neurotransmitter metabolism in the central nervous system. *J. Nutr.* **2001**, *131*, 846. [CrossRef]
51. Izumi, T.; Kawamura, K.; Ueda, H.; Bungo, T. Central administration of leucine, but not isoleucine and valine, stimulates feeding behavior in neonatal chicks. *Neurosci. Lett.* **2004**, *354*, 166–168. [CrossRef]
52. Wang, T.; Yao, W.; He, Q.; Shao, Y.; Zheng, R.; Huang, F. L-Leucine stimulates glutamate dehydrogenase activity and glutamate synthesis by regulating mTORC1/SIRT4 pathway in pig liver. *Anim. Nutr.* **2018**, *4*, 329–337. [CrossRef]

53. Gupta, A. Metabolism of Proteins and Amino Acids. In *Comprehensive Biochemistry for Dentistry*; Springer: Singapore, 2019; pp. 377–393.
54. Meister, A.; Anderson, M.E. Glutathione. *Annu. Rev. Biochem.* **1983**, *52*, 711–760. [CrossRef]
55. Yamane, H.; Suenaga, R.; Han, L.; Hayamizu, K.; Denbow, D.M.; Furuse, M. Intracerebroventricular injection of glutathione related dipeptides induce sedative and hypnotic effect under an acute stress in neonatal chicks. *Lett. Drug Des. Discov.* **2007**, *4*, 368–372. [CrossRef]
56. Yamane, H.; Tomonaga, S.; Suenaga, R.; Denbow, D.M.; Furuse, M. Intracerebroventricular injection of glutathione and its derivative induces sedative and hypnotic effects under an acute stress in neonatal chicks. *Neurosci. Lett.* **2007**, *418*, 87–91. [CrossRef] [PubMed]
57. Yamane, H.; Tsuneyoshi, Y.; Denbow, D.; Furuse, M. *N*-Methyl-*D*-aspartate and α-amino-3-hydroxy-5-methyl-4-isoxazolepropionate receptors involved in the induction of sedative effects under an acute stress in neonatal chicks. *Amino Acids* **2009**, *37*, 733–739. [CrossRef]
58. Zhu, H.; Blake, S.; Chan, K.T.; Pearson, R.B.; Kang, J. Cystathionine β-Synthase in Physiology and Cancer. *Biomed. Res. Int.* **2018**, *2018*, 3205125. [CrossRef] [PubMed]
59. Tran, P.V.; Tamura, Y.; Pham, C.V.; Elhussiny, M.Z.; Han, G.; Chowdhury, V.S.; Furuse, M. Neuropeptide Y modifies a part of diencephalic catecholamine but not indolamine metabolism in chicks depending on feeding status. *Neuropeptides* **2021**, *89*, 102169. [CrossRef]
60. Fernstrom, J.D.; Fernstrom, M.H. Tyrosine, phenylalanine, and catecholamine synthesis and function in the brain. *J. Nutr.* **2007**, *137*, 1539S–1547S. [CrossRef] [PubMed]
61. Parildar-Karpuzoğlu, H.; Doğru-Abbasoğlu, S.; Balkan, J.; Aykaç-Toker, G.; Uysal, M. Decreases in taurine levels induced by beta-alanine treatment did not affect the susceptibility of tissues to lipid peroxidation. *Amino Acids* **2007**, *32*, 115–119. [CrossRef]
62. Takeuchi, K.; Toyohara, H.; Sakaguchi, M. A hyperosmotic stress-induced mRNA of carp cell encodes Na(+)- and Cl(-)-dependent high affinity taurine transporter. *Biochim. Biophys. Acta* **2000**, *1464*, 219–230. [CrossRef]
63. Mujahid, A.; Furuse, M. Central administration of corticotropin-releasing factor induces thermogenesis by changes in mitochondrial bioenergetics in neonatal chicks. *Neuroscience* **2008**, *155*, 845–851. [CrossRef]
64. Davis, J.L.; Masuoka, D.T.; Gerbrandt, L.K.; Cherkin, A. Autoradiographic distribution of L-proline in chicks after intracerebral injection. *Physiol. Behav.* **1979**, *22*, 693–695. [CrossRef]
65. Tachibana, T.; Sato, M.; Oikawa, D.; Takahashi, H.; Boswell, T.; Furuse, M. Intracerebroventricular injection of neuropeptide Y modifies carbohydrate and lipid metabolism in chicks. *Regul. Pept.* **2006**, *136*, 1–8. [CrossRef] [PubMed]
66. Van Luijtelaar, E.L.; Van der Grinten, C.P.; Blokhuis, H.J.; Coenen, A.M. Sleep in the domestic hen (*Gallus domesticus*). *Physiol. Behav.* **1987**, *41*, 409–414. [CrossRef]
67. Chowdhury, V.S.; Shigemura, A.; Erwan, E.; Ito, K.; Bahry, M.A.; Tran, P.V.; Furuse, M. Oral administration of L-citrulline, but not L-arginine or L-ornithine, acts as a hypothermic agent in chicks. *J. Poult. Sci.* **2015**, *52*, 331–335. [CrossRef]
68. Kuenzel, W.J.; Masson, M. *A Stereotaxic Atlas of the Brain of the Chick (Gallus domesticus)*; The Johns Hopkins University Press: Baltimore, MD, USA, 1988.
69. Boogers, I.; Plugge, W.; Stokkermans, Y.Q.; Duchateau, A.L. Ultra-performance liquid chromatographic analysis of amino acids in protein hydrolysates using an automated pre-column derivatisation method. *J. Chromatogr. A* **2008**, *1189*, 406–409. [CrossRef]
70. Ito, K.; Bahry, M.A.; Hui, Y.; Furuse, M.; Chowdhury, V.S. Acute heat stress up-regulates neuropeptide Y precursor mRNA expression and alters brain and plasma concentrations of free amino acids in chicks. *Comp. Biochem. Physiol. A Mol. Integr. Physiol.* **2015**, *187*, 13–19. [CrossRef] [PubMed]
71. Kobayashi, K.; Pillai, K.S. Transformation of data and outliers. In *A Handbook of Applied Statistics in Pharmacology*; CRC Press, Taylor & Francis Group: New York, NY, USA, 2013; pp. 37–46.

 metabolites

MDPI

Article

Involvement of TauT/SLC6A6 in Taurine Transport at the Blood–Testis Barrier

Yoshiyuki Kubo [1,2,*,†], Sakiko Ishizuka [1,†], Takeru Ito [1], Daisuke Yoneyama [1], Shin-ichi Akanuma [1] and Ken-ichi Hosoya [1]

1 Department of Pharmaceutics, Graduate School of Medicine and Pharmaceutical Sciences, University of Toyama, 2630 Sugitani, Toyama 930-0194, Japan; s1860308@ems.u-toyama.ac.jp (S.I.); d2162302@ems.u-toyama.ac.jp (T.I.); m2061216@ems.u-toyama.ac.jp (D.Y.); akanumas@pha.u-toyama.ac.jp (S.A.); hosoyak@pha.u-toyama.ac.jp (K.H.)
2 Laboratory of Drug Disposition & Pharmacokinetics, Faculty of Pharma-Sciences, Teikyo University, Kaga 2-11-1, Tokyo 173-8605, Japan
* Correspondence: kubo.yoshiyuki.jf@teikyo-u.ac.jp; Tel.: +81-3-3964-8248
† Kubo Y. and Ishizuka S. contribute equally to this work.

Abstract: Taurine transport was investigated at the blood–testis barrier (BTB) formed by Sertoli cells. An integration plot analysis of mice showed the apparent influx permeability clearance of [^{3}H]taurine (27.7 μL/(min·g testis)), which was much higher than that of a non-permeable paracellular marker, suggesting blood-to-testis transport of taurine, which may involve a facilitative taurine transport system at the BTB. A mouse Sertoli cell line, TM4 cells, showed temperature- and concentration-dependent [^{3}H]taurine uptake with a K_m of 13.5 μM, suggesting that the influx transport of taurine at the BTB involves a carrier-mediated process. [^{3}H]Taurine uptake by TM4 cells was significantly reduced by the substrates of taurine transporter (TauT/SLC6A6), such as β-alanine, hypotaurine, γ-aminobutyric acid (GABA), and guanidinoacetic acid (GAA), with no significant effect shown by L-alanine, probenecid, and L-leucine. In addition, the concentration-dependent inhibition of [^{3}H]taurine uptake revealed an IC_{50} of 378 μM for GABA. Protein expression of TauT in the testis, seminiferous tubules, and TM4 cells was confirmed by Western blot analysis and immunohistochemistry by means of anti-TauT antibodies, and knockdown of TauT showed significantly decreased [^{3}H]taurine uptake by TM4 cells. These results suggest the involvement of TauT in the transport of taurine at the BTB.

Keywords: taurine; transport; blood-testis barrier; seminiferous tubules; antioxidant; infertility

Citation: Kubo, Y.; Ishizuka, S.; Ito, T.; Yoneyama, D.; Akanuma, S.; Hosoya, K. Involvement of TauT/SLC6A6 in Taurine Transport at the Blood–Testis Barrier. *Metabolites* **2022**, *12*, 66. https://doi.org/10.3390/metabo12010066

Academic Editors: Teruo Miyazaki, Takashi Ito, Alessia Baseggio Conrado and Shigeru Murakami

Received: 30 November 2021
Accepted: 8 January 2022
Published: 12 January 2022

1. Introduction

Taurine (2-aminoethanesulfonic acid) is known as a β-amino acid that abundantly exists in the human body, and its involvement in various physiological events, such as neuroprotection and osmotic regulation, has been suggested by cumulative studies [1–6]. In the biosynthesis of taurine, cysteine sulfonate decarboxylase (CSD) is assumed to be a rate-limiting enzyme, the activity of which is relatively low in humans [7,8], and experiments in rodents have revealed that a dietary deficiency of taurine results in a dramatic decrease of taurine in the liver [9], suggesting the importance of the taurine transport system for regulating the concentration of taurine in organs and tissues [1,10,11].

Taurine is also abundant in human semen, the concentration of which (679 μM) is reported to be maintained at levels 10-fold higher than in the blood, suggesting the importance of taurine in the testis, including germ cells [12]. In particular, low expression of antioxidant enzymes such as superoxide dismutase (SOD) in the testis is known, and abnormalities in the motility and form of sperm are caused by oxidative stress [13]. The oral administration of taurine showed a protective effect against arsenic-induced oxidative stress in rats, supporting the significant role of taurine in the testis [14], and the administration of taurine in streptozotocin-induced diabetic rats suggested that taurine acts as an antioxidant in the seminiferous tubules harboring germ cells [15]. The study in diabetic rats also supported the contribution of taurine to the seminiferous tubules, since its dietary intake improved the sperm characteristics, including sperm count and motility, that are closely related to male infertility [16–18].

These pieces of evidence indicate the relevance of taurine to the healthy maintenance and pathogeneses of sperm, suggesting the importance of regulating the concentration of such antioxidants in the seminiferous tubules, where Sertoli cells form the blood–testis barrier (BTB) to separate germ cells from the circulating blood [19,20]. At the BTB, Sertoli cells are assumed to form tight junctions to suppress non-specific transport via the paracellular route [21] and have been reported to express various transporter molecules, such as glucose transporters (GLUTs/SLC2As), L-type amino acid transporter 1 (LAT1/SLC7A5), monocarboxylate transporter 8 (MCT8/SLC16A2), sodium-dependent vitamin C transporters (SVCTs/SLC23As), concentrative nucleoside transporters (SLC28As), equilibrative nucleoside transporters (SLC29As), P-glycoprotein (P-gp/MDR1/ABCB1), and multidrug resistance-associated protein 1 (MRP1/ABCC1), involved in the transport of nutrients, metabolites, and xenobiotics at the BTB [22–27], suggesting the involvement of certain transporters in regulating taurine concentration in the seminiferous tubules.

Molecular cloning has achieved the identification of transporter molecules, and mouse taurine transporter (TauT/SLC6A6) and mouse γ-aminobutyric acid (GABA) transporter 3 (GAT3/SLC6A13, the orthologue of rat GAT2) mediate Na^+- and Cl^--dependent transport of taurine with K_m values of 4.50 and 540 μM, respectively [28–30]. In addition, H^+-coupled amino acid transporter 1 (PAT1/SLC36A1) was reported to have a low affinity for taurine (K_m = 7.5 mM) [31]. In particular, a study on the blood-to-retina and blood-to-liver transport of taurine reported the major contribution of TauT and rat GAT2 to taurine transport at the inner blood–retina barrier (inner BRB) and the hepatocytes, respectively, revealing precise mechanisms for regulating taurine concentration in the body [10,11].

However, little is known about the transport of taurine at the BTB, and this mechanism was investigated in the present study. The blood-to-testis transport of [^{3}H]taurine was investigated using integration plot analysis, and transport at the BTB was analyzed using a mouse-derived Sertoli cell line, TM4 cells [32]. In addition, the mRNA and protein expression of responsible transporters was analyzed using specific primers and antibodies.

2. Results

2.1. Integration Plot of [^{3}H]taurine in Mice

An integration plot analysis was carried out to evaluate the apparent influx clearance from circulating blood to the testis ($CL_{in,\ testis}$) (Figure 1), and the results obtained show that the $CL_{in,\ testis}$ of [^{3}H]taurine was 27.7 ± 0.2 μL/(min·g testis), which is approximately 3-fold greater than the $CL_{in,\ testis}$ of [^{14}C]D-mannitol (8.79 ± 3.58 μL/(min·g testis)).

2.2. Uptake of [^{3}H]taurine by TM4 Cells

An uptake analysis of [^{3}H]taurine was carried out in TM4 cells, where a time-dependent increase was shown in [^{3}H]taurine uptake for at least 20 min, with an initial uptake rate of 10.7 ± 0.2 μL/(min·mg protein) (Figure 2A). In addition, TM4 cells showed a significant reduction of [^{3}H]taurine uptake at 4 °C, and the uptake was also significantly decreased in the assay with Na^+-free, Cl^--free, and K^+-replacement buffers, with no change shown by extracellular pH (Figure 2B). The uptake of [^{3}H]taurine by TM4 cells took place in a concentration-dependent manner with K_m of 13.5 ± 3.8 μM, V_{max} of 3.42 ± 0.29 nmol/(min·mg protein), and K_d of 12.3 ± 0.65 μL/(min·mg protein), and the contribution ratio of the saturable process was calculated at approximately 95% (Figure 2C).

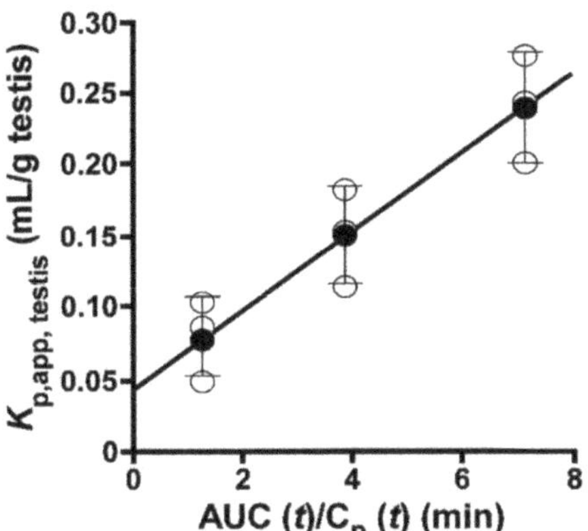

Figure 1. Blood-to-testis transport of [^{3}H]taurine. Initial uptake of [^{3}H]taurine by mouse testis. In integration plot analysis, [^{3}H]taurine (0.33 μM, 2 μCi/mouse) was injected into the left common carotid vein. Clearance was obtained from the regression line slope (shown as a solid line). The solid line was fitted using a nonlinear least-squares regression analysis program (MULTI). Each open circle represents an individual data point, and each closed circle represents mean ± standard deviation (SD) (n = 3).

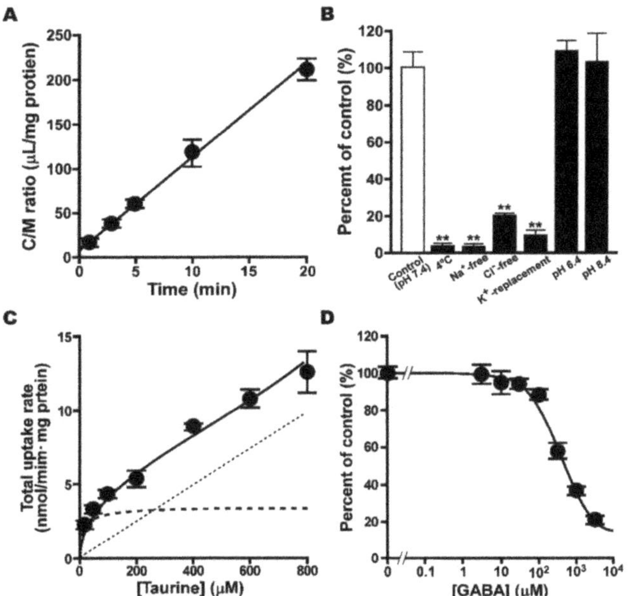

Figure 2. Uptake of [^{3}H]taurine by TM4 cells. (**A**) Time course of [^{3}H]taurine (16.5 nM, 0.1 μCi/well) uptake by TM4 cells was examined at 37 °C. (**B**) Effect of Na^+, Cl^-, membrane potential and extracellular pH on uptake was examined. Temperature dependence of uptake was examined at 4 °C for 5 min. (**C**) Concentration dependence of [^{3}H]taurine uptake was examined over a concentration range of 20-800 μM. Dashed, dotted, and solid lines represent saturable, non-saturable, and overall uptake of taurine, respectively. (**D**) Inhibitory effect of GABA on uptake was examined in the absence or presence of unlabeled GABA at designated concentrations. Inhibitory effect of GABA was evaluated as IC_{50} value (378 μM). Unless otherwise noted, uptake was examined at 37 °C for 5 min. Each point or column represents mean ± SD (n = 3). ** $p < 0.01$, significantly different from control.

The inhibition of [^{3}H]taurine uptake was also examined in TM4 cells, and uptake was significantly decreased in the presence of taurine, β-alanine, hypotaurine, GABA, and guanidinoacetic acid (GAA). No effect was observed in the presence of L-alanine, probenecid, and L-leucine (Table 1). In addition, the concentration-dependent inhibition of [^{3}H]taurine uptake was examined for GABA, and its IC_{50} was calculated as 378 μM (Figure 2D).

Table 1. Effect of several compounds on [^{3}H]taurine uptake by TM4 cells.

Compounds	Percentage of Control (%)		
Control	100	±	8
Taurine	3.70	±	0.26 **
β-Alanine	4.69	±	0.76 **
Hypotaurine	13.3	±	3.1 **
GABA	23.2	±	2.7 **
GAA	57.3	±	4.2 **
L-Alanine	82.1	±	2.0
Probenecid	98.7	±	3.9
L-Leucine	112	±	7

[^{3}H]Taurine (16.5 nM, 0.1 μCi/well) uptake was performed at 37 °C for 5 min in the absence (control) or presence of test compounds (1 mM). Each value represents mean ± SD (n = 3–9). ** $p < 0.01$, significantly different from control. GABA, γ-aminobutylic acid; GAA, guanidinoacetic acid.

2.3. Expression Analysis of TauT in TM4 Cells

mRNA expression of TauT was examined, and agarose gel electrophoresis clearly detected the PCR product for mouse TauT (189 bp) in mouse brain used as a positive control (Figure 3A). The product was also detected in mouse testis and TM4 cells (Figure 3A). For the analysis of protein expression, anti-TauT polyclonal antibodies were prepared by immunizing female Hartley guinea pigs with the C-terminus peptide of rat TauT. The specificity and cross-reactivity of anti-TauT polyclonal antibodies were confirmed by Western blot analysis to specifically detect the signal of glycosylated (75 kDa) and non-glycosylated (50 kDa) TauT proteins in mouse kidney (Figure 3B), and another Western blot analysis with the antibodies clearly detected the signals for TauT protein in mouse testis and TM4 cells (Figure 3C).

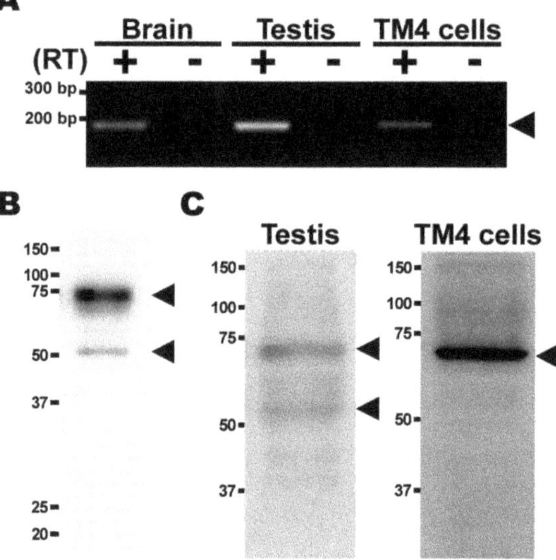

Figure 3. Expression study of TauT mRNA and protein. (**A**) mRNA expression of TauT in mouse brain, testis, and TM4 cells was analyzed by RT-PCR in the presence (+) or absence (−) of reverse

transcriptase (RT). PCR products were analyzed by agarose gel electrophoresis and visualized by staining with ethidium bromide. Arrowheads indicate predicted product sizes. (**B**) Specificity and cross-reactivity of anti-TauT antibody were confirmed by Western blot analysis. Mouse TauT protein was detected at approximately 75 kDa (glycosylated) and 50 kDa (non-glycosylated) (arrowhead). (**C**) Protein expression of TauT in mouse testis and TM4 cells was analyzed by Western blot analysis with anti-TauT antibody. Mouse TauT protein at approximately 50 kDa and 70 kDa (arrowhead).

Furthermore, the antibodies were recruited for the immunohistochemistry of TauT protein in mouse testis, and the fluorescence signal of TauT protein was detected in the seminiferous tubules (Figure 4A). In the study with anti-LAT1 antibodies, a strong signal merge of LAT1 and TauT was observed in the inner region of mouse seminiferous tubules during their weak merge in the outer region (Figure 4B) [26,30].

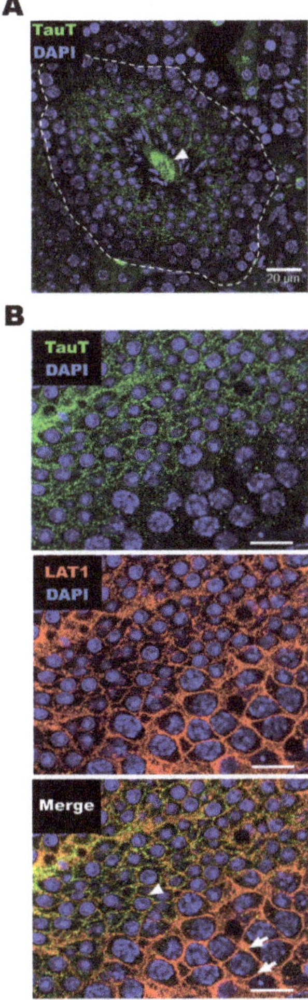

Figure 4. Immunohistochemistry of TauT in the mouse testis. (**A**) Mouse testis was analyzed with anti-TauT antibodies (green) and DNA intercalants DAPI (blue). TauT immunoreactivity was observed in seminiferous tubules (inside dotted line) and sperm (arrowhead). (**B**) Mouse seminiferous tubules were analyzed with anti-TauT antibody (green), anti-LAT1 antibody (red), and DNA intercalants

DAPI (blue). Signal was observed on basal (arrow) and lumen (arrowheads) sides of seminiferous tubules. Scale bar: 20 μm.

2.4. Knockdown of TauT in TM4 Cells

The effect of small interfering RNA specific for mouse TauT was examined in TM4 cells. Cells transfected with mouse TauT-specific siRNA showed a significant reduction of [^{3}H]taurine uptake by 53% (Figure 5A), and also significantly exhibited reduced mouse TauT transcript and protein expression by 81 and 57%, respectively (Figure 5B,C).

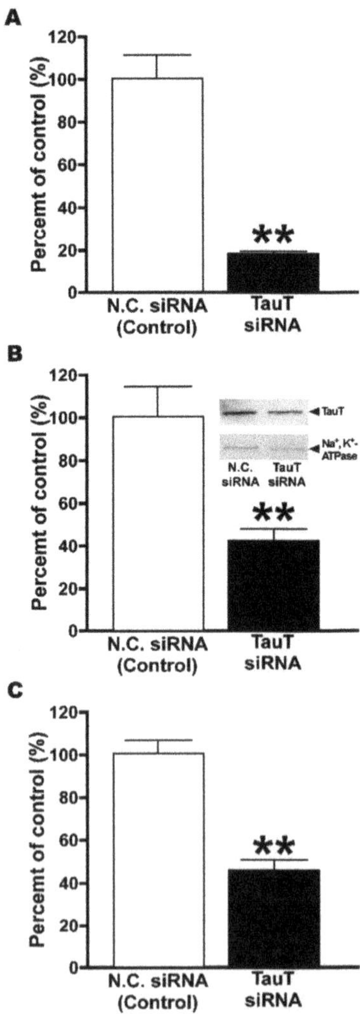

Figure 5. Knockdown analysis of TauT in TM4 cells. (**A**) Expression of TauT RNA was analyzed by quantitative real-time PCR and normalized with that of β-actin. (**B**) Expression of TauT protein was analyzed by Western blot with anti-TauT antibodies. Anti-Na^+/K^+-ATPase α1 antibodies were used to normalize the signal intensity of TauT (inset). (**C**) [^{3}H]Taurine (16.5 nM, 0.1 μCi/well) uptake was performed at 37 °C for 5 min. TM4 cells were treated with negative control siRNA (N.C. siRNA) or TauT siRNA. Each column represents mean ± SD (n = 3). ** $p < 0.01$, significantly different from control.

3. Discussion

The abundance of taurine in human semen implicated the importance of taurine in the testis [12]. Studies on male infertility have reported the protective effect of taurine against oxidative stress, showing improvement in sperm characteristics, such as sperm count and motility [16–18], suggesting the significance of taurine regulation in the seminiferous tubules harboring germ cells. Therefore, the involvement of certain transport systems in taurine transport at the BTB formed by Sertoli cells can be assumed.

In the integration plot analysis, the $CL_{in,testis}$ of [^{3}H]taurine was 27.7 μL/(min·g testis), which was much higher than that of paracellular transport marker (Figure 1). This suggests the blood-to-testis transport of taurine, which may involve a facilitative transport system at the BTB.

The calculation of $CL_{in,testis}$ for [^{3}H]taurine implies the involvement of taurine transport at the BTB, and uptake analysis with TM4 cells, a mouse Sertoli cell line, suggests the involvement of a carrier-mediated process in the transport of taurine at the BTB, since the uptake of [^{3}H]taurine took place in a time-, temperature-, and concentration-dependent manner (Figure 2). The estimation of kinetic parameters revealed a K_m value of 13.5 μM for taurine transport at the BTB, and further estimation of V_{max} (3.42 nmol/(min·mg protein)) and K_d (12.3 μL/(min·mg protein)) supported a substantial contribution of a saturable process, with a contribution ratio of 95%. In mice, taurine has been reported to be recognized by several transporters, such as TauT, mGAT3, and PAT1, and the K_m value (13.5 μM) obtained in TM4 cells was similar to that reported for mouse TauT (4.50 μM) [32]. In addition, the significant decrease in [^{3}H]taurine uptake by TM4 cells was shown in Na^+-free, Cl^--free, and K^+-replacement buffers, while no significant change was exhibited by different extracellular pH levels (Figure 2). These results support the possible contribution of TauT and mGAT3 to the transport of taurine at the BTB, since they are known as Na^+- and Cl^--dependent transporters during the pH-dependence of PAT1 [33].

In addition, the results obtained in the in vitro inhibition study support the possible contribution of TauT and mGAT3, since the uptake of [^{3}H]taurine by TM4 cells was significantly reduced in the presence of their substrates alanine, β-hypotaurine, GABA, and GAA [1,28,34], while it was not changed in the presence of L-alanine, a substrate of PAT1 (Table 1) [33]. Furthermore, the study of concentration-dependent inhibition suggests a major contribution of TauT to taurine transport at the BTB, since the calculated IC_{50} for GABA (378 μM) in TM4 cells was very similar to that of TauT (501 μM) (Figure 2D) [35], whereas mGAT3 is reported to transport GABA and taurine with a K_m of 18.0 and 540 μM, respectively [30].

The expression analysis suggests that TauT has a certain role in the BTB, and RT-PCR and Western blot analysis clearly detected mouse TauT mRNA and protein in testis and TM4 cells (Figure 3), clearly suggesting the specificity and cross-reactivity of anti-TauT antibodies, the epitope of which is the C-terminus peptide of rat TauT. In the immunohistochemistry with the antibodies, during the difficulty of identifying the type of cells, the detected signals supported the expression of TauT protein in the seminiferous tubules at least, and the plasma membrane localization of TauT protein was also suggested in the seminiferous tubules, since the merging of LAT1 and TauT was observed (Figure 4).

Furthermore, knockdown analysis of TM4 cells clearly supports the major contribution of mTauT to taurine transport at the BTB, since the transfection of siRNA designed for mouse TauT showed significantly decreased taurine transport with significantly reduced expression of mouse TauT mRNA and protein, revealing a contribution ratio of 93% for TauT (Figure 5).

4. Materials and Methods

4.1. Reagents, Animals and Cells

Commercially available chemicals of reagent grade were used in the present study. [2-^{3}H]Taurine ([^{3}H]taurine, 30.0 Ci/mmol) and [1-^{14}C]D-mannitol ([^{14}C]D-mannitol, 58 mCi/mmol) were purchased from American Radiolabeled Chemicals (St. Louis,

MO, USA). TM4 cells, a mouse-derived Sertoli cell line, were purchased from the UK Health Security Agency's European Collection of Authenticated Cell Cultures (Salisbury, UK). Horse serum and fetal bovine serum for culturing TM4 cells were purchased from Thermo Fisher Scientific (Waltham, MA, USA) and SAFC biosciences (Lenexa, KS, USA), respectively. Male ddY mice (8 weeks old) and female Hartley guinea pigs were purchased from Japan SLC (Hamamatsu, Japan) and were used in accordance with the guidelines for animal experiments (University of Toyama; registration #A2020PHA-4 and -5).

4.2. *Integration Plot Analysis*

Integration plot analysis was used to evaluate the apparent influx clearance ($CL_{in,\ testis}$) of [^{3}H]taurine from the circulating blood to the testis [36]. Briefly, after anesthetizing mice with an intraperitoneal injection of pentobarbital (6.48 mg/kg), [^{3}H]taurine (2 μCi/mouse) in 200 μL buffer (141 mM NaCl, 4.0 mM KCl, 10 mM 2-[4-(2-hydroxyethyl)piperazin-1-yl]ethanesulfonic acid (HEPES), and 2.8 mM $CaCl_2$) was injected into the internal jugular vein. Blood samples were collected at designated times, and mice were decapitated to collect the testes, which were lysed in 2N NaOH. Plasma samples were prepared by centrifugation (5000× *g*, 4 °C, 10 min) of collected blood samples, and radioactivity in the blood, plasma, and testis samples was determined by means of a liquid scintillation counter (LSC-7400, Hitachi Healthcare Manufacturing, Chiba, Japan). In the present study, [^{3}H]D-mannitol (1 μCi/mouse) was also tested as a non-permeable paracellular transport marker. $CL_{in,testis}$ was calculated by Equation (1):

$$V_d = CL_{in,testis} \times AUC(t)/C_p(t) + V_i \quad (1)$$

where the apparent volume of distribution to the testis, the area under the plasma concentration time curve of the compound from time 0 to t, the plasma concentration of the compound at time t, and the rapidly equilibrated distribution volume of the compound in the testis are expressed as V_d (μL/g testis), AUC(t) (dpm min/mL), C_p(t) (dpm/mL), and V_i (μL/g testis), respectively.

4.3. *Cell Uptake Analysis*

TM4 cells were cultured on TrueLine cell culture dishes (NIPPON Genetics, Tokyo, Japan) in a humidified environment at 37 °C and 5% CO_2. For culturing TM4 cells, Dulbecco's modified Eagle medium (D-MEM)/Ham's F-12 medium with L-glutamine (FUJIFILM Wako Pure Chemical Corporation, Osaka, Japan) was supplemented with 100 U/mL benzylpenicillin, 100 μg/mL streptomycin sulfate, 10% horse serum, and 5% fetal bovine serum [32]. The medium was changed every 3–4 days, and trypsin-EDTA was used for cell passage.

In the uptake study, TM4 cells (1.0×10^5 cells/well) were seeded on Corning BioCoat poly-D-lysine-coated 24-well clear flat bottom TC-treated multiwell plates (Corning, Corning, NY, USA). After culturing for 2 days, the cells were rinsed 3 times with extracellular fluid (ECF) buffer (122 mM NaCl, 25 mM $NaHCO_3$, 3 mM KCl, 1.2 mM $MgSO_4$, 0.4 mM K_2HPO_4, 10 mM HEPES, 10 mM D-glucose, 1.4 mM $CaCl_2$, pH 7.4) warmed to 37 °C, and incubated in 200 μL ECF buffer containing [^{3}H]taurine (0.1 μCi/well) at 37 °C. As reported elsewhere [37], radioactivity measurements were performed using a liquid scintillation counter (LSC-7400, Hitachi Healthcare Manufacturing) after the cells were lysed with 1N NaOH and neutralized with 1N HCl. [^{3}H]Taurine uptake by TM4 cells was expressed by the cell-to-medium (C/M) ratio (μL/mg protein) obtained in Equation (2), and cellular protein content was determined using a DC protein assay kit and microplate reader (Model 680) purchased from Bio-Rad (Hercules, CA, USA)

$$\text{Cell-to-medium ratio} = ([^3\text{H}]\ \text{dpm per cell protein (mg)})/([^3\text{H}]\ \text{dpm per μL ECF-buffer}) \quad (2)$$

As described previously [36], in the study of concentration dependence, kinetic parameters were obtained by a nonlinear least-squares regression analysis program (MULTI) [38] and Equation (3):

$$V = V_{max} \times C/(K_m + C) + K_d \times C \quad (3)$$

where C, V, K_m, V_{max}, and K_d are the substrate concentration, the uptake rate, the Michaelis constant, and the maximal uptake rate and non-saturable uptake clearance, respectively.

MULTI was also used to calculate the half maximal inhibitory concentration (IC_{50}) by Equation (4) [37]:

$$P = (P_{max} - P_{min})/[(1 + (I/IC_{50})^n] + P_{min} \quad (4)$$

where P and P_{max} are the relative [^{3}H]taurine uptake (%) in the presence or absence of γ-aminobutyric acid (GABA), respectively, and P_{min}, n, and I are the inhibitor-insensitive component of relative [^{3}H]taurine uptake with GABA, the Hill coefficient, and the concentration of GABA, respectively.

4.4. *mRNA Expression Analysis*

Total RNA extraction and mRNA expression analysis using reverse transcription polymerase chain reaction (RT-PCR) were carried out as reported previously [39]. Briefly, the total RNA of TM4 cells was extracted using an RNeasy Micro Kit (Qiagen, Venlo, Netherlands) and ReverTraAce (TOYOBO, Osaka, Japan), and oligo dT primer was used for reverse transcription. Veriti Thermal Cycle (Thermo Fisher Scientific, Waltham, MA, USA) and Ex-Taq (Takara, Shiga, Japan) were used in the PCR for mouse TauT through 30 cycles at 94 °C for 30 s, 57 °C for 30 s, and 72 °C for 45 s. The nucleic acid alignment of primers for mouse TauT (NM_009320) was as follows: forward primer was 5′-GCGTTTCCCGTACCTCTGC-3′ and reverse primer was 5′-ATGGATGCGTAGCCAATGCC-3′. Amplified products in PCR were subjected to agarose gel electrophoresis, followed by ultraviolet visualization with ethidium bromide.

In quantitative real-time PCR, Mx3005P (Agilent Technologies, Santa Clara, CA, USA), SYBR Premix Ex Taq, and ROX reference dye (Takara, Shiga, Japan) were used through 35 cycles at 95 °C for 30 s, 57 °C for 30 s, and 72 °C for 30 s [39]. The nucleic acid alignment of primers was as follows: forward primer was 5′-GCGTTTCCCGTACCTCTGC-3′ and reverse primer was 5′-ATGGATGCGTAGCCAATGCC-3′ for mouse TauT (NM_009320), and forward primer was 5′-TCATGAAGTGTGACGTTGACATCCGT-3′ and reverse primer was 5′-CCTAGAAGCATTTGCGGTGCACGATG-3′ for β-actin (NM_031144.3). The initial amount of mouse TauT transcripts was estimated by determining the threshold cycle number, and a standard curve was prepared by the control plasmids harboring the gene fragment of mouse TauT.

4.5. *Protein Expression Analysis*

Anti-TauT polyclonal antibodies were prepared by using the epitope encompassing 30 amino acid residues (REGATPFHSRATLMNGALMKPSHVIVETMM) located at the C-terminus region of rat TauT (NP_058902). Male Hartley guinea pigs were immunized with the glutathione S-transferase (GST)-tagged epitope peptide, and the titer and specificity of antibodies were improved in affinity column purifications after blood serum was collected. Western blot analysis was performed as described elsewhere [36]. Briefly, the crude membrane fraction (20 μg protein) prepared from TM4 cells was analyzed, with anti-TauT polyclonal antibodies (1.0 μg/mL) and horseradish peroxidase-conjugated anti-guinea pig IgG antibodies (Merck Millipore, Burlington, MA, USA) used as primary and secondary antibodies, respectively. In the analysis of Na^+/K^+-ATPase α-1 subunit, a plasma membrane marker, anti-Na^+/K^+-ATPase α-1 antibodies (0.5 μg/mL; Merck Millipore, Burlington, MA, USA) and horseradish peroxidase-conjugated anti-mouse IgG antibodies (Merck Millipore) were used. ECL Prime Western Blotting Detection System (Merck Millipore, Burlington, MA, USA) and Luminescent Image Analyzer (LAS-4000, FUJIFILM) were used for signal detection.

Immunohistochemistry was performed as reported previously [36], and an LSM780 confocal microscope (Carl Zeiss, Oberkochen, Germany) was used. After tissue fixation with 4% paraformaldehyde, a cryostat (CM1900, Leica, Wetzlar, Germany) was used to prepare frozen sections (12 μm thickness) of ddY mice testes, which were mounted on glass slides for blocking with 10% goat serum (Nichirei, Tokyo, Japan) at room temperature. Anti-TauT polyclonal antibodies (2.0 μg/mL) and rabbit polyclonal anti-LAT1 antibodies (3.0 μg/mL; Trans Genic, Fukuoka, Japan) were applied as the primary antibodies. Alexa Fluor 488-conjugated goat anti-guinea pig IgG and CyTM3-conjugated donkey anti-rabbit IgG antibodies were obtained from Thermo Fisher Scientific and Jackson ImmunoResearch, respectively, and were used as secondary antibodies. After being treated with 4′,6-diamidino-2-phenylindole (DAPI) and VECTASHIELD mounting medium (Vector Laboratories, Burlingame, CA, USA), sections were examined by confocal microscopy.

4.6. Knockdown Analysis of TauT

Small interfering RNA (siRNA) of mouse TauT were designed with reference to previous reports. Stealth RNAi Negative Control Medium GC Duplexes were obtained from Thermo Fisher Scientific [39], and all processes of transfection to TM4 cells were carried out according to the manufacturer's instructions [39]. In brief, siRNA (30 pmol/well) was introduced to TM4 cells cultured on 6-well plates by means of Lipofectamine RNAiMAX (5 μL/well; Thermo Fisher Scientific). The uptake study was initiated 48 h after transfection, and the knockdown efficiency was confirmed by Western blot analysis and quantitative real-time PCR.

4.7. Statistical Analysis

All data in this paper are expressed as mean ± SD. Statistically significant differences (p-Value < 1%) between two or more than three groups were determined by unpaired two-tailed Student's t-test or one-way ANOVA by Dunnett's test, respectively.

5. Conclusions

Clinically, hormonal modulators and herbal medicines have been used in the treatment of male infertility [40]. It is also thought that taurine is promising in the treatment of male infertility, since it has a protective effect against oxidative stress in the testes to improve sperm characteristics [14–18] and has been assumed not to produce adverse effects [41]. In the present study, the involvement of a transport system in the blood-to-testis transport of taurine is suggested, implying a facilitative transport system for taurine at the BTB. Uptake and expression studies have suggested that TauT largely contributes to the influx transport of taurine at the BTB, and it is assumed that TauT has an important role in the protection of germ cells from oxidative stress by supplying taurine to seminiferous tubules. The present findings will be helpful for improving the treatment of male infertility in association with a better understanding of taurine regulation in the testis.

Author Contributions: Y.K., S.A. and K.H.: conception and design; S.I., T.I. and D.Y.: validation, collection, and assembly of data; S.I., T.I. and Y.K.: data analysis and interpretation; Y.K. and K.H.: writing and editing manuscript. All authors have read and agreed to the published version of the manuscript.

Funding: The present study was supported in part by Grant-in-Aids from the Japan Society for Promotion of Science, such as Scientific Research (C) (KAKENHI: 20K07173), and Research Grants from the Tamura Science and Technology Foundation.

Institutional Review Board Statement: The animal study protocol was approved by University of Toyama (#A2020PHA-4 and A2020PHA-5).

Informed Consent Statement: Not applicable.

Data Availability Statement: The data of this study are available from the corresponding author upon reasonable request.

Conflicts of Interest: The authors declare that they have no conflict of interest.

References

1. Kubo, Y.; Akanuma, S.-I.; Hosoya, K.-I. Impact of SLC6A Transporters in Physiological Taurine Transport at the Blood–Retinal Barrier and in the Liver. *Biol. Pharm. Bull.* **2016**, *39*, 1903–1911. [CrossRef]
2. Kaplan, B.; Dinçer, S.; Babül, A.; Duyar, I. The effect of taurine administration on vitamin C levels of several tissues in mice. *Amino Acids* **2003**, *27*, 225–228. [CrossRef]
3. Das, J.; Ghosh, J.; Manna, P.; Sil, P. Taurine provides antioxidant defense against NaF-induced cytotoxicity in murine hepatocytes. *Pathophysiology* **2008**, *15*, 181–190. [CrossRef] [PubMed]
4. Lambert, I.H. Regulation of the cellular content of the organic osmolyte taurine in mammalian cells. *Neurochem. Res.* **2004**, *29*, 27–63. [CrossRef]
5. Takatani, T.; Takahashi, K.; Uozumi, Y.; Shikata, E.; Yamamoto, Y.; Ito, T.; Matsuda, T.; Schaffer, S.; Fujio, Y.; Azuma, J. Taurine increases testicular function in aged rats by inhibiting oxidative stress and apoptosis. *Amino Acids.* **2015**, *47*, 1549–1558.
6. Yu, J.; Kim, K. Effect of taurine on antioxidant enzyme system in B16F10 melanoma cells. *Adv. Exp. Med. Biol.* **2009**, *7*, 491–499.
7. Yuan, L.-Q.; Xie, H.; Luo, X.-H.; Wu, X.-P.; Zhou, H.-D.; Lu, Y.; Liao, E.-Y. Taurine transporter is expressed in osteoblasts. *Amino Acids* **2006**, *31*, 157–163. [CrossRef]
8. Hayes, K.C.; Stephan, Z.F.; Sturman, J.A. Growth Depression in Taurine-Depleted Infant Monkeys. *J. Nutr.* **1980**, *110*, 2058–2064. [CrossRef] [PubMed]
9. Chen, W.; Matuda, K.; Nishimura, N.; Yokogoshi, H. The effect of taurine on cholesterol degradation in mice fed a high-cholesterol diet. *Life Sci.* **2004**, *74*, 1889–1898. [CrossRef]
10. Tomi, M.; Terayama, T.; Isobe, T.; Egami, F.; Morito, A.; Kurachi, M.; Ohtsuki, S.; Kang, Y.-S.; Terasaki, T.; Hosoya, K.-I. Function and regulation of taurine transport at the inner blood–retinal barrier. *Microvasc. Res.* **2007**, *73*, 100–106. [CrossRef]
11. Ikeda, S.; Tachikawa, M.; Akanuma, S.-I.; Fujinawa, J.; Hosoya, K.-I. Involvement of γ-aminobutyric acid transporter 2 in the hepatic uptake of taurine in rats. *Am. J. Physiol. Liver Physiol.* **2012**, *303*, G291–G297. [CrossRef] [PubMed]
12. Holmes, R.P.; Goodman, H.; Shihabi, Z.K.; Jarow, J.P. The taurine and hypotaurine content of human semen. *J. Androl.* **1992**, *13*, 289–292. [PubMed]
13. Alahmar, A.T. Role of Oxidative Stress in Male Infertility: An Updated Review. *J. Hum. Reprod. Sci.* **2019**, *12*, 4–18. [CrossRef]
14. Das, J.; Ghosh, J.; Manna, P.; Sinha, M.; Sil, P.C. Taurine protects rat testes against $NaAsO_2$-induced oxidative stress and apoptosis via mitochondrial dependent and independent pathways. *Toxicol. Lett.* **2009**, *187*, 201–210. [CrossRef]
15. Tsounapi, P.; Saito, M.; Dimitriadis, F.; Koukos, S.; Shimizu, S.; Satoh, K.; Takenaka, A.; Sofikitis, N. Antioxidant treatment with edaravone or taurine ameliorates diabetes-induced testicular dysfunction in the rat. *Mol. Cell. Biochem.* **2012**, *369*, 195–204. [CrossRef]
16. Mohamed, N.; Gawad, H.A. Taurine dietary supplementation attenuates brain, thyroid, testicular disturbances and oxidative stress in streptozotocin-induced diabetes mellitus in male rats. *Beni-Suef Univ. J. Basic Appl. Sci.* **2017**, *6*, 247–252. [CrossRef]
17. Lanzafame, F.M.; La Vignera, S.; Vicari, E.; Calogero, A. Oxidative stress and medical antioxidant treatment in male infertility. *Reprod. Biomed. Online* **2009**, *19*, 638–659. [CrossRef]
18. Terai, K.; Horie, S.; Fukuhara, S.; Miyagawa, Y.; Kobayashi, K.; Tsujimura, A. Combination therapy with antioxidants improves total motile sperm counts: A Preliminary Study. *Reprod. Med. Biol.* **2019**, *19*, 89–94. [CrossRef]
19. Cheng, C.Y.; Mruk, D.D. The Blood-Testis Barrier and Its Implications for Male Contraception. *Pharmacol. Rev.* **2011**, *64*, 16–64. [CrossRef] [PubMed]
20. Lui, W.Y.; Mruk, D.; Lee, W.M.; Cheng, C.Y. Sertoli Cell Tight Junction Dynamics: Their Regulation During Spermatogenesis1. *Biol. Reprod.* **2003**, *68*, 1087–1097. [CrossRef]
21. Kerr, J.B. Ultrastructure of the seminiferous epithelium and intertubular tissue of the human testis. *J. Electron. Microsc. Tech.* **1991**, *19*, 215–240. [CrossRef]
22. Angulo, C.; Maldonado, R.; Pulgar, E.; Mancilla, H.; Cordova, A.; Villarroel, F.; Castro, M.A.; Concha, I.I. Vitamin C and oxidative stress in the seminiferous epithelium. *Biol. Res.* **2011**, *44*, 169–180. [CrossRef]
23. Bae, H.S.; Jin, Y.K.; Ham, S.; Kim, H.K.; Shin, H.; Cho, G.; Lee, K.J.; Lee, H.; Kim, K.M.; Koo, O.J.; et al. CRISRP/Cas9-mediated knockout of Mct8 reveals a functional involvement of Mct8 in testis and sperm development in a rat. *Sci. Rep.* **2020**, *10*, 11148. [CrossRef] [PubMed]
24. Bart, J.; Hollema, H.; Groen, H.J.M.; Veries, E.G.E.; Hendrikse, N.H.; Sleijfer, D.T.; Wegman, T.D.; Vaalburg, W.; Graaf, W.T.A. The distribution of drug-efflux pumps, P-gp, BCRP, MRP1 andMRP2, in the normal blood–testis barrier and in primary testicular tumours. *Eur. J. Cancer.* **2004**, *40*, 2064–2070. [CrossRef] [PubMed]
25. Huang, S.; Ye, J.; Yu, J.; Chen, L.; Zhou, L.; Wang, H.; Li, Z.; Wang, C. The accumulation and efflux of lead partly depend on ATP-dependent efflux pump–multidrug resistance protein 1 and glutathione in testis Sertoli cells. *Toxicol. Lett.* **2014**, *226*, 277–284. [CrossRef] [PubMed]
26. Kato, R.; Maeda, T.; Akaike, T.; Tamai, I. Nucleoside Transport at the Blood-Testis Barrier Studied with Primary-Cultured Sertoli Cells. *J. Pharmacol. Exp. Ther.* **2004**, *312*, 601–608. [CrossRef]

27. Nakada, N.; Mikami, T.; Hana, K.; Ichinoe, M.; Yanagisawa, N.; Yoshida, T.; Endou, H.; Okayasu, I. Unique and selective expression of L-amino acid transporter 1 in human tissue as well as being an aspect of oncofetal protein. *Histol. Histopathol.* **2013**, *29*, 2.
28. Smith, K.; Borden, L.; Wang, C.H.; Hartig, P.R.; Branchek, T.; Weinshank, R.L. Cloning and expression of a high affinity taurine transporter from rat brain. *Mol. Pharmacol.* **1992**, *42*, 563–569.
29. Liu, Q.R.; López-Corcuera, B.; Nelson, H.; Mandiyan, S.; Nelson, N. Cloning and expression of a cDNA encoding the transporter of taurine and/8-alanine in mouse brain. *Proc. Natl. Acad. Sci. USA* **1992**, *89*, 12145–12149. [CrossRef]
30. Liu, Q.; López-Corcuera, B.; Mandiyan, S.; Nelson, H.; Nelson, N. Molecular characterization of four pharmacologically distinct gamma-aminobutyric acid transporters in mouse brain [corrected]. *J. Biol. Chem.* **1993**, *268*, 2106–2112. [CrossRef]
31. Anderson, C.M.; Grenade, D.S.; Boll, M.; Foltz, M.; Wake, K.A.; Kennedy, D.J.; Munck, L.K.; Miyauchi, S.; Taylor, P.M.; Campbell, F.C.; et al. H+/amino acid transporter 1 (PAT1) is the imino acid carrier: An intestinal nutrient/drug transporter in human and rat. *Gastroenterology* **2004**, *127*, 1410–1422. [CrossRef]
32. Mather, J.P. Establishment and characterization of two distinct mouse testicular epithelial cell lines. *Biol. Reprod.* **1980**, *23*, 243–252.
33. Metzner, L.; Neubert, K.; Brandsch, M. Substrate specificity of the amino acid transporter PAT1. *Amino Acids* **2006**, *31*, 111–117. [CrossRef]
34. Tachikawa, M.; Kasai, Y.; Yokoyama, R.; Fujinawa, J.; Ganapathy, V.; Terasaki, T.; Hosoya, K.-I. The blood-brain barrier transport and cerebral distribution of guanidinoacetate in rats: Involvement of creatine and taurine transporters. *J. Neurochem.* **2009**, *111*, 499–509. [CrossRef] [PubMed]
35. Vinnakota, S.; Qian, X.; Egal, H.; Sarthy, V.; Sarkar, H.K. Molecular characterization and in situ localization of a mouse retinal taurine transporter. *J. Neurochem.* **2002**, *69*, 2238–2250. [CrossRef] [PubMed]
36. Kubo, Y.; Obata, A.; Akanuma, S.-I.; Hosoya, K.-I. Impact of Cationic Amino Acid Transporter 1 on Blood-Retinal Barrier Transport of L-Ornithine. *Investig. Opthalmology Vis. Sci.* **2015**, *56*, 5925. [CrossRef]
37. Kubo, Y.; Shimizu, Y.; Kusagawa, Y.; Akanuma, S.-I.; Hosoya, K.-I. Propranolol Transport Across the Inner Blood–Retinal Barrier: Potential Involvement of a Novel Organic Cation Transporter. *J. Pharm. Sci.* **2013**, *102*, 3332–3342. [CrossRef]
38. Yamaoka, K.; Tanigawara, Y.; Nakagawa, T.; Uno, T. A pharmacokinetic analysis program (multi) for microcomputer. *J. Pharm. -Dyn.* **1981**, *4*, 879–885. [CrossRef]
39. Kubo, Y.; Yahata, S.; Miki, S.; Akanuma, S.-I.; Hosoya, K.-I. Blood-to-retina transport of riboflavin via RFVTs at the inner blood-retinal barrier. *Drug Metab. Pharmacokinet.* **2017**, *32*, 92–99. [CrossRef] [PubMed]
40. Haidl, G.; Schill, W.-B. Guidelines for Drug Treatment of Male Infertility. *Drugs* **1991**, *41*, 60–68. [CrossRef] [PubMed]
41. Azuma, J.; Sawamura, A.; Awata, N. Usefulness of taurine in chronic congestive heart failure and its prospective application: Current therapy of intractable heart failure. *Jpn Circ. J.* **1992**, *56*, 95–99. [CrossRef]

 metabolites

Article

Taurine Ameliorates Streptozotocin-Induced Diabetes by Modulating Hepatic Glucose Metabolism and Oxidative Stress in Mice

Shigeru Murakami [1,*], Kohei Funahashi [1], Natsuki Tamagawa [1], Ma Ning [2] and Takashi Ito [1]

1 Department of Bioscience and Biotechnology, Fukui Prefectural University, 4-1-1 Matsuoka Kenjyojima, Eiheiji 910-1195, Fukui, Japan; kouhei19950314@gmail.com (K.F.); s1321027@g.fpu.ac.jp (N.T.); tito@fpu.ac.jp (T.I.)

2 Division of Health Science, Graduate School of Health Science, Suzuka University of Medical Science, 1001-1, Kishioka, Suzuka 510-0293, Mie, Japan; maning@suzuka-u.ac.jp

* Correspondence: murakami@fpu.ac.jp

Abstract: Taurine is a sulfated amino acid derivative that plays an important role in maintaining the cell function of the living body. Although taurine has been shown to ameliorate diabetes, its mechanism of action has not yet been fully elucidated. The present study investigated the effects of taurine on diabetes focusing on glucose metabolism and oxidative stress. Type 1 diabetes was induced by the administration of streptozotocin (STZ) to male C57BL/6J mice. Taurine was dissolved in drinking water at 3% (*w/v*) and allowed to be freely ingested by diabetic mice. The weight and blood glucose levels were measured weekly. After nine weeks, mice were sacrificed and their serum, liver, and kidney were removed and used for biochemical and histological analyses. A microarray analysis was also performed in normal mice. Taurine alleviated STZ-induced hyperglycemia and hyperketonemia, accompanied by the suppression of the decrease in hepatic glycogen and upregulation of the mRNA expression of hepatic glucose transporter GLUT-2. Furthermore, STZ-induced elevation of oxidative stress in the liver and kidney was suppressed by taurine treatment. These results showed that taurine ameliorated diabetes and diabetic complications by improving hepatic glucose metabolism and reducing oxidative stress.

Keywords: diabetes; taurine; GLUT-2; oxidative stress; glycogen; glucose metabolism

Citation: Murakami, S.; Funahashi, K.; Tamagawa, N.; Ning, M.; Ito, T. Taurine Ameliorates Streptozotocin-Induced Diabetes by Modulating Hepatic Glucose Metabolism and Oxidative Stress in Mice. *Metabolites* **2022**, *12*, 524. https://doi.org/10.3390/metabo12060524

Academic Editor: Cholsoon Jang

Received: 15 April 2022
Accepted: 31 May 2022
Published: 6 June 2022

1. Introduction

Taurine, a sulfur-containing amino acid derivative, is abundant in the body of mammals and plays a pivotal role in maintaining cellular homeostasis, such as through osmotic pressure regulation, protein stabilization, antioxidant and anti-inflammatory actions, and calcium ion regulation [1,2]. Taurine is also involved in basic vital activities such as fetal development [3], the maintenance of the organ function, and metabolism [4–8]. Many experiments using animal models and cultured cells have demonstrated that taurine supplementation can suppress the development of a wide range of diseases [9–13]. The important role of taurine in the body has also been demonstrated in taurine transporter knockout mice, in which the tissue taurine concentration is markedly decreased due to the inhibition of the cellular taurine uptake [14]. These mice exhibit a deteriorated organ function and accelerated aging [15–17].

Among the many disease-preventing effects of taurine, anti-diabetic effects have been reported in animal models and humans [18,19]. In streptozotocin- or alloxan-induced type 1 diabetic animal models, taurine suppresses the development of diabetes [20] and diabetic complications [21,22]. Anti-diabetic effects of taurine have also been demonstrated in type 2 diabetic models, including fructose- or high-fat diet-induced animals and Otsuka Long-Evans Tokushima Fatty (OLETF) rats [23,24].

Taurine exhibits hypoglycemic, insulin-sensitizing, and insulin secretagogue activities. Although most of the beneficial effects of taurine on type 1 diabetes have been attributed to its protective effect on pancreatic β-cells, other potential actions have been postulated as anti-diabetic mechanisms of taurine. Taurine regulates the K_{ATP} channel and enhances K^+-induced depolarization in pancreatic β-cells, which leads to increased insulin secretion [25]. Taurine also increases the islet Ca^{2+} uptake in response to glucose and stimulates insulin release [26]. In addition, taurine elevates extracellular glucose concentrations thorough the glucose transporter GLUT-2 in β-cells and thereby increases insulin secretion [27]. Thus, taurine directly affects pancreatic β-cells and stimulates insulin secretion.

In addition to the direct effects on pancreatic β-cells, taurine has been shown to affect the insulin target organs, such as liver and skeletal muscle, and ameliorate hyperglycemia and insulin resistance. Taurine improves fatty acid-induced hepatic insulin resistance by modulating the insulin signaling pathway in rats [28]. In the liver and skeletal muscle, taurine enhances the glucose uptake, which is related to amelioration of hyperglycemia and insulin resistance [29]. In vitro and in vivo studies suggest that the hypoglycemic effects of taurine may be mediated through the interaction of taurine with insulin receptor [30]. Taurine supplementation increases both basal and insulin-stimulated tyrosine phosphorylation of the insulin receptor in skeletal muscle and liver [27]. In obese ob/ob mice, taurine improves insulin resistance by activating AMP-activated protein kinase (AMPK) in skeletal muscle [31].

Thus, several mechanisms are considered to be involved in anti-diabetic effects. The liver plays a central role in glucose metabolism and is important in the development of diabetes, but few studies have investigated the action of taurine. In this study, we focused on the glucose metabolism and oxidative stress and investigated the effect of taurine on the development of diabetes.

2. Results

2.1. Body Weight and Blood Levels of Glucose, Insulin, and Ketone Body

Control mice gained body weight steadily, but the body weight gain of STZ-treated diabetic mice was completely suppressed (Figure 1). Taurine treatment had no significant effect on food intake and body weight throughout the experimental period. The blood glucose levels in STZ diabetic mice increased until the third week, and then remained around 400 mg/dL (Figure 2A). Taurine treatment significantly suppressed the increase in blood glucose levels at the first, second, and fifth weeks. In other weeks, blood glucose levels were also lower in taurine-treated mice than in diabetic mice, but the difference was not significant. Serum insulin levels were markedly decreased in STZ diabetic mice (Figure 2B). Taurine treatment partially recovered the decrease in insulin levels, but the effect was not significant. Blood levels of ketone body were doubled in STZ mice compared with those of control mice (Figure 2C), and taurine treatment significantly suppressed the elevation of blood ketone body levels.

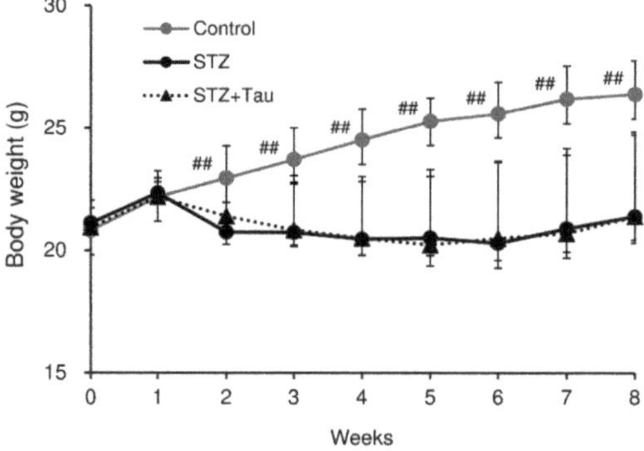

Figure 1. Changes in body weight in control, diabetic, and taurine-treated groups. Each value is expressed as the mean ± S.D. (n = 8–12), ## $p < 0.01$ vs. STZ group.

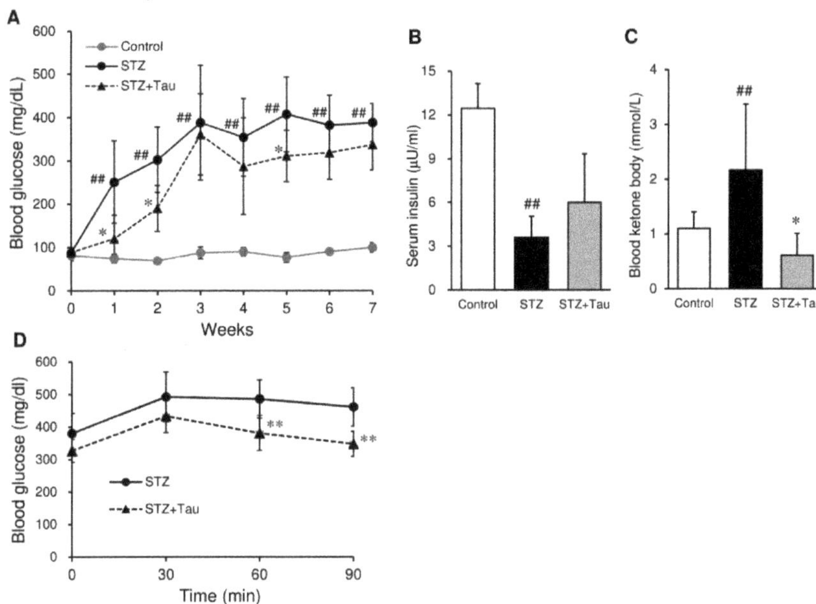

Figure 2. Effects of taurine supplementation on the levels of blood glucose, insulin, and ketone body as well as insulin resistance in diabetic mice. Blood glucose was measured weekly after blood was drawn from the tail vein (**A**). Serum levels of insulin (**B**) and ketone body (**C**) were measured at week 8. Insulin resistance was assessed using an intraperitoneal glucose tolerance test at week 7 (**D**). Each value is expressed as the mean ± S.D. (n = 7–12), ## $p < 0.01$ vs. Control group, * $p < 0.05$, ** $p < 0.01$ vs. STZ group.

2.2. Insulin Resistance

A glucose tolerance test was performed to evaluate the effect of taurine treatment on insulin resistance. Glucose challenge elevated the blood glucose levels, which remained high in STZ mice (Figure 2D). Blood glucose levels of taurine-treated mice were significantly lower than those in the STZ group at 60 and 90 min after glucose loading.

2.3. Glycogen Levels of the Liver and Kidney

Glucose and glycogen content were measured in the liver and kidney. The amounts of glucose and glycogen in the liver of STZ mice were significantly lower than those of control mice (Figure 3A,B). In contrast, the glycogen content in the kidney was significantly increased by STZ administration (Figure 3C). These STZ-induced alterations of glycogen and glucose content were significantly recovered by taurine treatment.

2.4. MRNA Expression of Glucose Metabolism-Related Genes in the Liver

To examine the effect of taurine on glucose metabolism, the expression of glucose and lipid metabolism-related genes was investigated in the liver. STZ mice showed a higher mRNA expression of phosphoenolpyruvate carboxykinase (PEPCK) and sterol regulatory element-binding protein 1c (SREBP-1c) than control mice (Figure 4A,E), but a lower mRNA expression of glucokinase (GK) and glucose-6-phosphatase (G6P) (Figure 4B,C). Although the expression of glucose transporter-2 (GLUT-2) was not affected by STZ, taurine treatment significantly increased the GLUT-2 mRNA expression (Figure 4D).

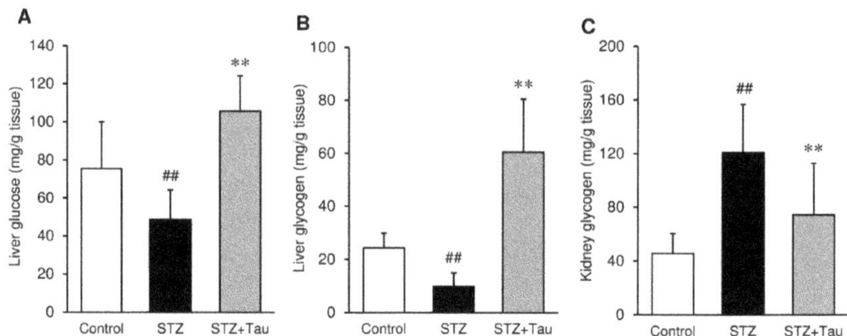

Figure 3. Effects of taurine supplementation on the glucose and glycogen content in the liver and kidney in diabetic mice. (**A**) Liver glucose, (**B**) Liver glycogen, (**C**) Kidney glycogen. Glucose and glycogen levels were measured using commercially available kits. Each value is expressed as the mean ± S.D. (n = 7–8). ## $p < 0.01$ vs Control group, ** $p < 0.01$ vs. STZ group.

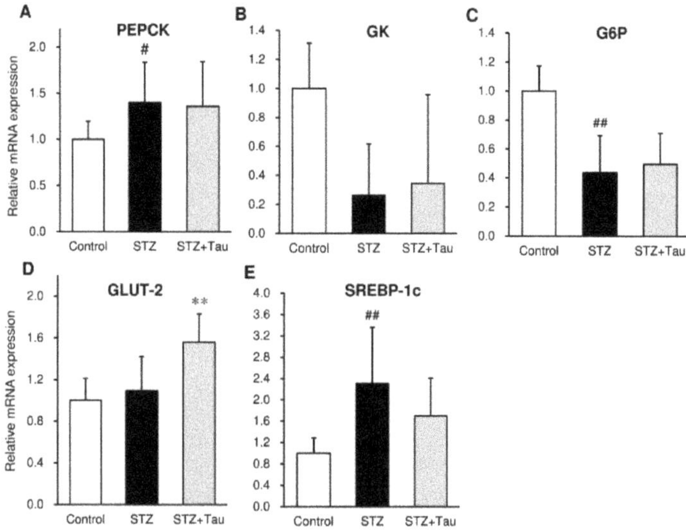

Figure 4. Effects of taurine supplementation on the mRNA expression of glucose metabolism-related genes in the liver of diabetic mice. Total RNA was extracted, and the mRNA expression was assessed using RT-PCR. (**A**) Phosphoenolpyruvate carboxykinase (PEPCK), (**B**) glucokinase (GK), (**C**) glucose-6-phosphatase (G6P), (**D**) glucose transporter-2 (GLUT-2), (**E**) sterol regulatory element-binding protein 1c (SREBP-1c). Each value is expressed as the mean ± S.D. (n = 4), # $p < 0.05$, ## $p < 0.01$ vs. Control group, ** $p < 0.01$ vs. STZ group.

2.5. Microarray Analyses

To further evaluate the effect of taurine treatment on the expression of genes responsible for glycogen accumulation in the liver, a microarray analysis was performed on normal mice. Initially, we focused on the differentially expressed genes (DEGs) defined as genes with a fold change of more than 1.5 or less than 0.67 in taurine-treated group compared to the control group and the p-value was less than 0.01 in the student's t-test (Figure 5A). A total of 38 genes were detected as DEGs (16 upregulated and 22 downregulated genes (Table 1). Based on the restrictive false discovery rate analysis (Benjamini and Hochberg's method), only one gene was listed as a differentially expressed gene (Slc25a38) in taurine-treated mice; the genes associated with glycogen deposition were not detected (Figure 5).

Therefore, we focused on the genes annotated with gene ontology terms for glycogen biosynthesis, glucose metabolic process, UDP-glucose metabolic process and glucose transport (Figure 5). Consistent with the gene expression analysis in STZ-induced diabetic mice, GLUT-2 (Slc2a2) was increased by 1.3-fold in taurine-treated mice. In addition, the expression of UDP-glucose pyrophosphorylase 2 (Ugp2), glycine N-methyltransferase (Gnmt), and GLUT-9 (Slc2a9) was increased, while the expression of the genes responsible for the regulation of the glucose metabolic process, such as transcription factors (Myc, Crem) and phosphatase (Ppp1r2), was decreased in the taurine-treated mice. These results suggest that upregulation of GLUT2 and UGP2 may be involved in taurine-induced glycogen increase in the liver (Figure 6).

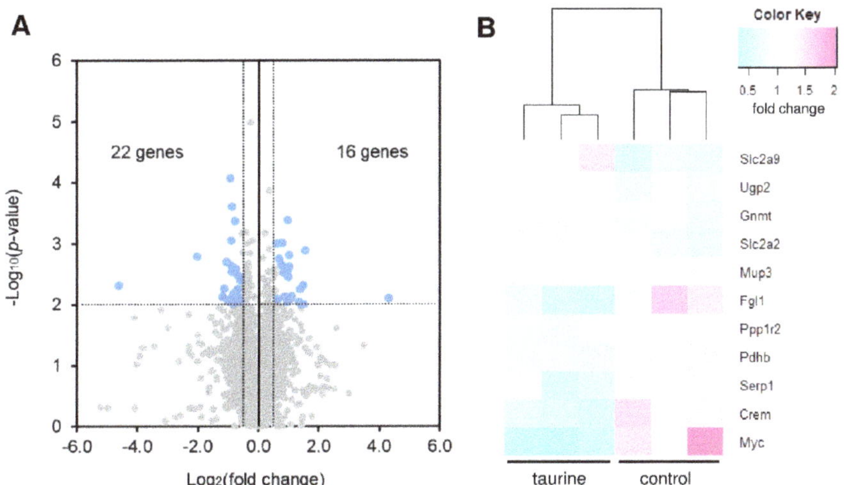

Figure 5. Microarray analysis in the liver from taurine-treated mice. (**A**) Volcano plots showing the differential expression patterns between taurine-treated and control groups. (**B**) Heat maps showing the differentially expressed genes ($p < 0.05$) annotated with GO term for glycogen biosynthesis (GO: 0005978), glucose metabolic process (GO: 0006006), UDP-glucose metabolic process (GO: 0006011) and glucose transport (GO: 0046323) in the liver between taurine-treated and control mice.

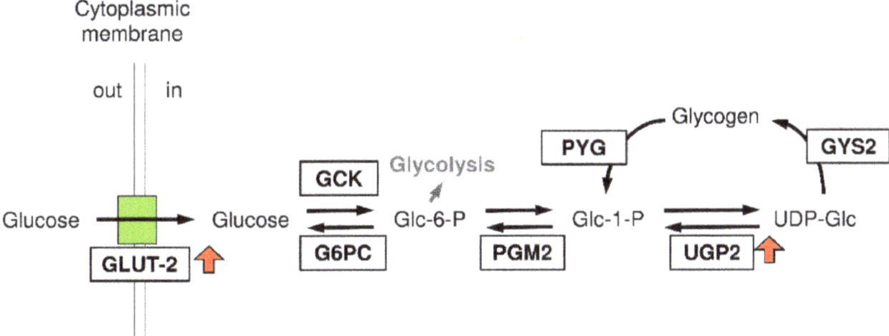

Figure 6. Cellular pathways of glycogen synthesis stimulation by taurine. G6PC: Glucose 6-phosphatase; GCK: Glucokinase; PGM2: Phosphoglucomutase 2; UGP2: UDP-glucose pyrophosphorylase 2; GYS2: Glycogen synthase 2; PYG: Glycogen phosphorylase.

Table 1. Differential expressed genes in the liver of taurine-treated mice identified by microarray analysis.

Gene Symbol	Gene Description	Ratio	*p*-Value
Mt2	metallothionein 2	19.662	0.0077
Aldh1b1	aldehyde dehydrogenase 1 family, member B1	2.893	0.0013
Slc16a7	solute carrier family 16, member 7	2.822	0.0099
Sucnr1	succinate receptor 1	2.767	0.0048
Cyp3a59	cytochrome P450, family 3, subfamily a, polypeptide 59	2.611	0.0097
Ddah1	dimethylarginine dimethylaminohydrolase 1	2.555	0.0055
Hmgn2	high mobility group nucleosomal binding domain 2	2.104	0.0072
Hnmt	histamine N-methyltransferase	2.073	0.0078
Sult1a1	sulfotransferase family 1A, phenol-preferring, member 1	2.025	0.0023
Cyp2j6	cytochrome P450, family 2, subfamily j, polypeptide 6	2.007	0.0015
Siae	sialic acid acetylesterase	1.986	0.0035
Manea	mannosidase, endo-alpha	1.942	0.0090
Papss2	3-phosphoadenosine 5-phosphosulfate synthase 2	1.932	0.0034
Slc6a12	solute carrier family 6	1.926	0.0026
Acss2	acyl-CoA synthetase short-chain family member 2	1.925	0.0004
Slc2a9	solute carrier family 2, member 9	1.816	0.0073
Ech1	enoyl coenzyme A hydratase 1, peroxisomal	0.657	0.0039
Rela	v-rel reticuloendotheliosis viral oncogene homolog A	0.644	0.0089
Ube2m	ubiquitin-conjugating enzyme E2M	0.627	0.0032
H2afz	H2A histone family, member Z	0.623	0.0073
Ldah	lipid droplet associated hydrolase	0.615	0.0055
Mrpl38	mitochondrial ribosomal protein L38	0.579	0.0004
Selk	selenoprotein K	0.578	0.0074
Nfyb	nuclear transcription factor-Y beta	0.576	0.0067
Gucd1	guanylyl cyclase domain containing 1	0.558	0.0066
Mtus1	mitochondrial tumor suppressor 1	0.551	0.0097
Usp6nl	USP6 N-terminal like	0.544	0.0002
Cd9	CD9 antigen	0.543	0.0028
Lurap1l	leucine rich adaptor protein 1-like	0.535	0.0008
Gpr182	G protein-coupled receptor 182	0.533	0.0078
Paqr9	progestin and adipoQ receptor family member IX	0.530	0.0023
Arpc1b	actin related protein 2/3 complex, subunit 1B	0.520	0.0001
Inhbc	inhibin beta-C	0.508	0.0087
Crem	cAMP responsive element modulator	0.482	0.0019
Txndc5	thioredoxin domain containing 5	0.453	0.0054
Gm10181	predicted gene 10181	0.442	0.0074
Myc	myelocytomatosis oncogene	0.247	0.0016
Csad	cysteine sulfinic acid decarboxylase	0.041	0.0048

2.6. Oxidative Stress in the Liver and Kidney

The effects of taurine supplementation on oxidative stress were determined in the liver and kidneys. The levels of malondialdehyde (MDA), a marker of lipid peroxidation, were elevated in the liver and kidneys of diabetic mice compared to those of normal mice (Figure 7A,B). Taurine supplementation suppressed the increase in MDA. Consistent with these results, staining with 8-hydroxy-2′-deoxyguanosine (8-OHdG) antibody, a marker of oxidative DNA damage, was increased in the liver and kidneys of STZ mice, but decreased in these organs upon taurine treatment (Figure 7C,D).

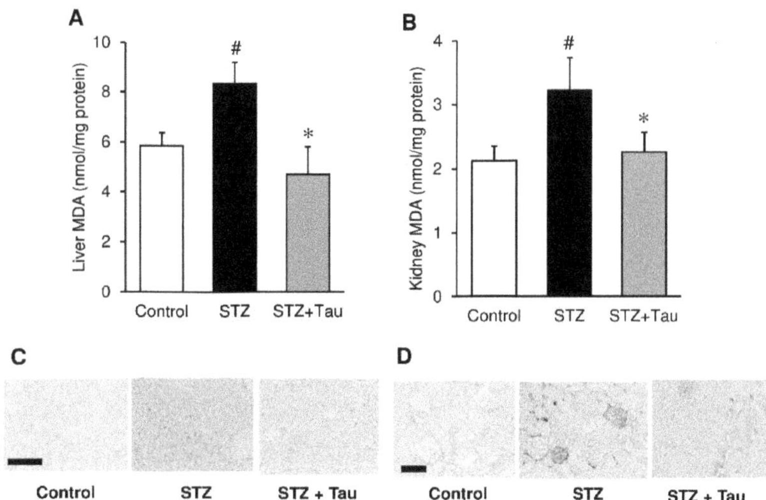

Figure 7. Effects of taurine supplementation on tissue oxidative stress in diabetic mice. MDA levels in liver (**A**) and kidney (**B**) were measured as an oxidative stress marker. Immunostaining 8-OHdG was performed in the liver (**C**) and kidney (**D**). Each value is expressed as the mean ± S.D. (n = 7–8), # $p < 0.05$ vs Control group, * $p < 0.05$ vs. STZ group. Scale bar = 100 μm.

2.7. Liver Taurine Content

Although the liver taurine content of diabetic mice was significantly lower than that of control mice, taurine treatment restored the reduction in taurine (Figure 8A). Immunohistochemical analyses using anti-taurine antibody supported this finding (Figure 8B).

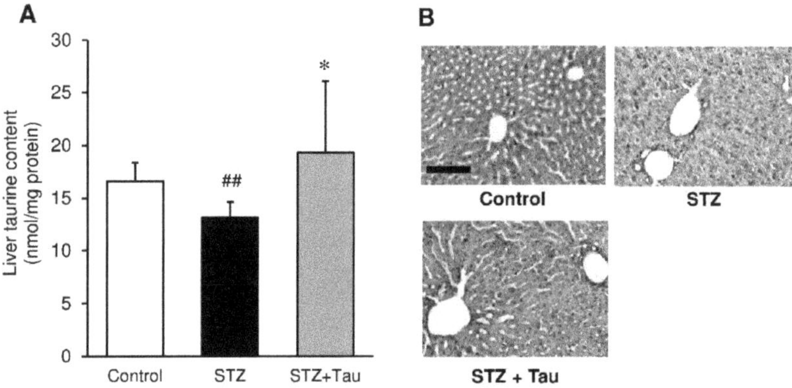

Figure 8. Effects of taurine supplementation on liver taurine content in diabetic mice. Taurine was extracted from the liver and then determined using HPLC (**A**). The liver tissue was fixed with formalin and stained with an anti-taurine antibody (**B**). Each value is expressed as the mean ± S.D. (n = 7–8), ## $p < 0.01$ vs Control group, * $p < 0.05$ vs STZ group. Scale bar = 100 μm.

2.8. Pancreatic Insulin and Taurine

Localization of taurine and insulin in the pancreas was analyzed using immunofluorescent double staining. The results showed colocalization of taurine (green) with insulin (red) in β-cells of pancreatic islet (Figure 9A). Pancreatic islets of STZ diabetic mice were characterized by a lower β-cell percentage and lower β-cell immunoreactivity for insulin than in control mice. The insulin area was also markedly reduced in STZ diabetic mice (Figure 9B). The intensity of insulin immunoreactivity and area of insulin were increased by taurine supplementation.

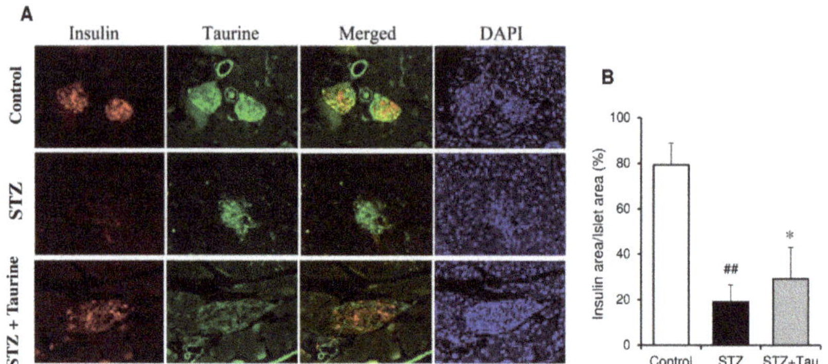

Figure 9. Effects of taurine supplementation on the insulin and taurine levels in pancreas of diabetic mice. Double immunofluorescence staining for taurine and insulin was performed in the pancreas (**A**). The areas of insulin-positive cells were measured, and the percentage in the islets was calculated (**B**). Each value is expressed as the mean ± S.D. (n = 24), ## $p < 0.01$ vs Control group, * $p < 0.05$ vs. STZ group.

3. Discussion

The present results showed that taurine treatment ameliorated hyperglycemia and insulin resistance in STZ-induced diabetic mice. The hypoglycemic effect of taurine was mild, and in some weeks, there were no significant differences in blood glucose compared with diabetic mice. There have been conflicting reports concerning the hypoglycemic effect of taurine, depending on the animal models used and the experimental conditions [32,33]. The present biochemical data of the liver and blood suggest that the anti-diabetic effect of taurine may be associated with improvement of hepatic glucose metabolism.

Glucose is liberated from dietary carbohydrate by hydrolysis within the small intestine and then absorbed into the blood. Elevated concentrations of glucose in the blood stimulate the release of insulin from pancreatic β-cells, and insulin acts on cells to stimulate the uptake, utilization, and storage of glucose, which results in a decrease in blood glucose levels. The liver plays a major role in the modulation of glucose homeostasis by controlling various pathways of glucose metabolism. Hepatic glucose production is primarily regulated by phosphoenolpyruvate carboxykinase (PEPCK) and glucose-6-phosphatase (G6P), which are the rate-limiting enzymes in gluconeogenesis [34]. An analysis of the hepatic mRNA expression related to glucose synthesis, including PEPCK and G6Pase, showed that taurine treatment had no significant effects on the mRNA expression of these enzymes, suggesting that gluconeogenesis is not responsible for the anti-diabetic effect of taurine. In contrast, the mRNA expression of hepatic GLUT-2 was increased by taurine treatment. GLUT-2 is the major type of glucose transporter responsible for the hepatic glucose uptake and utilization [35]. Glucose enters hepatocytes via GLUT-2 and is phosphorylated by glucokinase (GK) and then used to synthesize glycogen. A microarray analysis using normal mice showed that taurine increased the expression of GLUT-2 as well as UGP2, an enzyme involved in glycogen synthesis. These results indicate that taurine has the capacity to promote the uptake of glucose and glycogen synthesis in the liver. In type 1 diabetes, insulin deficiency reduces the hepatic glucose uptake, resulting in hepatic glucose deficiency. However, taurine can normalize hepatic glucose metabolism through the upregulation of the glucose uptake and glycogen synthesis in the liver.

Insulin secretion from β-cells is reduced in type I diabetes, leading to a deficiency of glucose and glycogen levels in the liver. As a result, the hepatic ketone body production from the oxidation of fatty acids increases and is used peripherally as an energy source, which results in diabetic ketoacidosis [36,37]. The levels of blood ketone body were twice as high in STZ diabetic mice as in normal mice in the present study. Consistent with the recovery of hepatic glycogen levels, blood ketone body levels were significantly reduced by taurine treatment. Thus, the suppression of serum ketone body levels by taurine reflects the improvement of glucose metabolism and glycogen content in the liver.

The hypoglycemic effect of taurine in type I diabetic models has been mainly explained by the β-cell protective action of the pancreas and the subsequent suppression of the decrease in insulin secretion [38,39]. The insulin levels of serum and pancreas were markedly decreased in STZ diabetic mice compared to normal mice. Taurine administration to STZ diabetic mice recovered serum insulin levels, but not to a significant degree. In contrast, a histochemical analysis indicated that taurine treatment significantly recovered the area of pancreatic insulin. These data indicate that taurine has a weak β-cell -protective effect. Therefore, improvement of hepatic glucose metabolism by taurine is mainly explained by its stimulatory action on the glucose uptake and subsequent glycogen synthesis in the liver, as shown above, but its protective effect on pancreatic β-cells may also be involved.

The administration of STZ or alloxane causes diabetic complications due to chronic hyperglycemia. It is well established that reactive oxygen species (ROS) generation or oxidative stress plays an important role in the onset and development of diabetic complications [40,41]. Taurine has been reported to attenuate diabetic complications in STZ- or alloxane-induced diabetic animals, including nephropathy, retinopathy, neuropathy, and cardiovascular problems [42,43]. Taurine has also been shown to suppress liver and kidney injuries caused by toxic substances, including ethanol [44], acetaminophen [45], and arsenic [46]. It is known that the protective effect of taurine on the development of diabetes, diabetic complications, and tissue injuries are closely associated with the reduction of oxidative stress. We previously showed that taurine treatment suppressed the high-fat-diet-induced reduction in the enzyme activity and molecules involved in hepatic anti-oxidant defense, including superoxide dismutase, catalase, and glutathione [47]. In type I diabetes, hepatic glycogen content is decreased due to deficiency of insulin. In contrast, kidney glycogen content is increased. While kidneys usually contain low amounts of glycogen, large deposits of glycogen are present in diabetic kidneys [48]. One of the hallmarks of renal damage in STZ diabetic mice is the accumulation of glycogen [49,50]. Liver glycogen is required for normal energy production, but excessive glycogen accumulation in the kidney impairs the renal function. In the present study, although the kidney glycogen content was elevated in STZ diabetic mice, taurine treatment suppressed this elevation of glycogen levels, suggesting a suppressive effect of taurine on diabetic complications. Consistent with previous findings, biochemical and histochemical data revealed that taurine reduced oxidative stress, as evidenced by the reduction in MDA content and immunostaining of 8-OHdG in both liver and kidney of diabetic mice. In addition, immunofluorescence staining of insulin in the pancreas indicated that taurine protected pancreatic β-cells from toxicity induced by STZ. The cytotoxic action of STZ is mediated by ROS. As oxidative stress is involved in the induction of insulin resistance [28] and β-cell dysfunction, the anti-oxidative action of taurine may be related to not only suppression of diabetic complications, but also amelioration of insulin resistance and hyperglycemia. It is known that hyperglycemia increases tissue oxidative stress by promoting ROS production, which activates NF-κB and induces cellular apoptosis [51]. In fact, taurine has been reported to suppress inflammation and apoptosis in a diabetic model, suggesting that taurine's anti-diabetic activity is associated with these effects in addition to its antioxidant activity [33].

The present study showed that liver taurine content was reduced in STZ diabetic mice compared to control mice. Immunofluorescence staining showed that pancreatic taurine was also reduced in STZ diabetic mice. Given the important role of taurine in the body, including its antioxidant and osmoregulatory actions, a decrease in taurine in the pancreas may be associated with pancreatic β-cell damage and decreased insulin secretion. In contrast, taurine supplementation to diabetic mice increased pancreatic taurine levels and restored insulin secretion. These findings suggest that taurine plays a role in protecting cells from oxidative damage and suppressing functional deterioration in diabetic mice. Similar reductions in tissue taurine levels have been previously reported in diabetic animals and humans. Indeed, plasma and platelet taurine levels were significantly lower in diabetic patients than in healthy subjects [52]. Taurine levels in the liver were reduced in alloxan-induced diabetic rats [53]. We previously showed that taurine treatment suppressed the high-fat-diet-induced reduction of the enzyme activity and molecules involved in hepatic anti-oxidant defense [47]. Based on these data, reduced liver taurine levels in STZ diabetic mice may be involved at least in part in the disturbance of the glucose metabolism and anti-oxidant defense system in the liver.

The present study and previously reported results suggest that compounds and food components with antioxidant and anti-inflammatory properties may be effective in preventing the progression of diabetes and diabetic complications. Some naturally occurring compounds, such as flavonoids, exhibit antidiabetic effects through their antioxidant and anti-inflammatory actions [54,55]. Thus, taurine is expected to be effective in preventing diabetes and other diseases caused by abnormal glucose metabolism.

4. Materials and Methods

4.1. Experimental Animals and Treatment

Six-week-old male C57BL/6J mice weighing 19–21 g were purchased from CLEA Japan Inc. (Tokyo, Japan) and housed under a 12-h light/dark cycle at 22 ± 2 °C and 50% ± 5% humidity. The mice had free access to food and water. After 1 week of acclimation, mice were randomly divided into the following three groups of 12 animals each: normal group with citrate buffer as a vehicle control, STZ-treated diabetic group (STZ) and STZ plus taurine group (STZ + Tau). Diabetic mice were given a single intraperitoneal injection of 200 mg/kg STZ (Sigma-Aldrich Inc., St. Louis, MO, USA). STZ was dissolved in citrate buffer at pH 4.5 and used within 20 min of preparation. Induction of diabetes was confirmed by measuring the glucose levels in blood obtained from the tail vein. Taurine was dissolved in drinking water at 3% (*w/v*) and allowed to be freely ingested by mice. Animals were allowed free access to diet and drinking water. The body weight and food intake were monitored every other day. The blood glucose levels were measured once a week using a blood glucometer Nipro Stat Strip (Nipro, Osaka, Japan). Nine weeks after STZ injection, mice were deprived of food overnight, and blood samples were withdrawn from the ophthalmic vein under a mixed anesthetic agent (0.3 mg/kg of medetomidine, 4.0 mg/kg of midazolam, and 5.0 mg/kg of butorphanol; Fujifilm Wako Pure Chemical Co., Osaka, Japan). Liver and kidney tissue were dissected out, weighed, frozen in liquid nitrogen, and stored −80 °C. All of procedures were performed in accordance with the ARRIVE guidelines.

4.2. Glucose Tolerance Teat

The mice were fasted overnight and were intraperitoneally injected with glucose solution (2 g/kg body weight). The blood samples were collected from the tail vein of the mice and glucose levels were measured at 0, 30, 90 min after injection using a blood glucometer Nipro Stat Strip (Nipro, Osaka, Japan).

4.3. Biochemical Measurements

Serum was obtained by centrifuging at 3000 rpm for 15 min at 4 °C. The serum insulin levels were determined using a commercial ELISA kit (Morinaga Institute of Biological Science, Yokohama, Japan). Blood ketone body concentrations were measured with a blood glucometer Nipro Stat Strip (Nipro) using a dedicated β-hydroxybutyrate test strip.

4.4. Glucose and Glycogen Determination of Liver and Kidney

Small pieces of the liver and kidney were homogenized with 10 times the amount of ice-cold 50 mM Tris-HCl buffer (pH 7.4). The homogenate was centrifuged at 15,000× g for 20 min at 4 °C, and the supernatant was used to measure glucose and glycogen levels. The glucose and glycogen content was determined using a commercially available kit; the Glucose Colorimetric/Fluorometric Assay Kit and Glycogen Colorimetric Assay Kit II (Bio Vison, Waltham, MA, USA), respectively, according to the manufacturer's instructions.

4.5. Malondialdehyde (MDA) Determination of Liver and Kidney

MDA was determined as a marker of lipid peroxidation. The supernatant of tissue homogenates as shown above was used to measure MDA levels. The amount of MDA in the liver and kidneys was determined using a commercially available kit; the Malondialdehyde Assay Kit (Nikken Seil Co., Ltd., Shizuoka, Japan) according to the manufacturer's instructions.

4.6. Liver Taurine Content Determination

The liver tissue was homogenized in 100 mM HEPES (pH 7.5), and 4 volumes of 5% sulfosalicylic acid were added to the tissue lysate. After centrifugation, the supernatant was filtered and neutralized with 1 M $NaHCO_3$. The samples were subjected to high-performance liquid chromatography (HPLC) to determine the taurine concentration, according to a previously reported method [56]. In brief, the supernatant was derivatized with an OPA reagent (3 mg of o-phthalaldehyde with 50 µL of 95% ethanol, 10 µL of 2-mercaptoethanol in 5 mL of 100 mM borate buffer, pH 10.4) and then subjected to HPLC (D-2000; Hitachi High Technologies, Tokyo, Japan) equipped with a reverse-phase column (Cosmosil 5C18-MS-II, 140 mm; Nacalai Tesque, Kyoto, Japan).

4.7. RNA Isolation and Real-Time Polymerase Chain Reaction (PCR)

Total RNA was extracted from liver tissues using Sepasol (Nacalai Tesque) according to the manufacturer's protocol. cDNA was generated from total RNA by reverse transcription with Rever Tra Ace (Toyobo, Osaka, Japan). Quantitative reverse transcription (RT)-PCR was performed using StepOne (Applied Biosystems, Waltham, MA, USA) with the Thunderbird SYBR qPCR Kit (Toyobo, Osaka, Japan).The primer sequences are listed in Table 2.

Table 2. Primer sequences uded in the experiment.

Gene	Forward Primer	Reverse Primer
GLUT-2	TTCATGTCGGTGGGACTTGTG	TGGCAGTCATGCTCACGTAACT
G-6-P	AACGTCTGTCTGTCCCGGATCTAC	TTCCGGAGGCTGGCATTGTA
GK	GTACGACCGGATGGTGGATG	TCTACCAGCTTGAGCAGCAC
PEPCK	CGAATGTGTGGGCGATGAC	ACTGAGGTGCCAGGAGCAACT
SREBP1c	AATGACAAGATTGTGGAGCTCAAAG	ACACCAGGTCCTTCAGTGATT

4.8. Histological and Immunohistochemical Studies

Liver and kidney tissue samples from mice were fixed overnight in 4% paraformaldehyde, followed by dehydration and paraffin infiltration. Embedding and sectioning procedures were then performed to construct 5-µm sections using Leica Microsystems (Wetzlar, Germany). The expression levels of taurine and 8-OHdG were determined in liver and kidneys, and that of 8-OHdG was additionally determined in kidneys by immunohistochemical (IHC) staining. Liver or kidney sections were subjected to dewaxing and rehydration, and antigen retrieval was then performed using a 500-W microwave for 5 min, after which a non-specific binding protein was blocked by incubation in 1% skim milk. After blocking endogenous peroxidase activity by incubation in 1% skim milk for 25 min at room temperature, sections were incubated overnight with a primary antibody mouse monoclonal anti-8-OHdG antibody (5 mg/mL, Japan Institute for the Control of Aging, Shizuoka, Japan) overnight at room temperature in a humid chamber. Following three washes in phosphate-buffered saline (PBS), the sections were subsequently treated with a specific biotinylated secondary antibody and avidin-biotin-peroxidase conjugate (Vector Laboratories, Burlingame, CA, USA). Immunoreaction was visualized by incubation with a

DAB peroxidase substrate kit (Nacalai Tesque) and counterstained with hematoxylin in taurine immunostaining for liver.

The expression levels of insulin and taurine was measured in pancreas tissue samples using double immunofluorescence staining technique. In brief, paraffin-embedded tissues were deparaffinized in xylene and rehydrated in serial alcohol solutions. Antigen retrieval was then performed using a 500-W microwave for 5 min, and non-specific binding protein was blocked by incubation in 1% skim milk. Tissue samples were stained using antibodies against insulin (I2018; SIGMA-ALDRICH, St Louis, MO, USA) and taurine at a 1:400 dilution, overnight. A rabbit polyclonal taurine antibody without a cross-reaction was produced as described previously [57]. After washing with PBS, donkey anti-mouse IgG Alexa Flour 594 secondary antibody (A11011, Invitrogen, Waltham, MA, USA) and donkey anti-rabbit IgG Alexa Flour 488 secondary antibody (A11012; Invitrogen, Waltham, MA, USA) at 1:400 dilution in PBS were incubated for 2 h. Stained tissues were mounted using DAPI-fluoromount-G (Southern Biotech, Birmingham, AL, USA) and observed under a fluorescence microscope (Olympus, Tokyo, Japan). The intensity was evaluated in four different areas of each sample using the ImageJ software program (NIH, Bethesda, MD, USA) for target molecules, and the intensity ratio was calculated in comparison to that of the nuclear staining of DAPI, as a reference for the adjustment of the cell number.

4.9. Microarray and Pathway Analyses

Six-week-old male ICR mice were treated with taurine for 8 weeks with 2% (*w/v*) taurine-containing drinking water. RNA from their livers was isolated using Sepazol-RNA Super G (Nacalai Tesque, Kyoto, Japan) and cleaned using an RNeasy Mini Kit (Qiagen GmbH, Hilden, Germany). A microarray analysis was performed on two groups (control and taurine-treated mice, $n = 3$ for each group) using Clariom S arrays (Affymetrix). The array was scanned by using GeneChip™ Scanner 3000 7G (Thermo Fisher Inc., Waltham, MA, USA) and image analysis was performed by using Expression Console™ Software (Thermo Fisher Inc., Waltham, MA, USA). These procedures were performed by Filgen Inc. (Nagoya, Japan). The microarray data are posited in the Gene Expression Omnibus.

4.10. Statistics

Results are presented as the means ± standard deviations (S.D.). Comparisons among groups were performed using an analysis of variance and coupled with Dunnett's tests. All p-values less than 0.05 were considered to be significant.

5. Conclusions

The present results revealed that taurine supplementation ameliorated hyperglycemia and insulin resistance in STZ-induced type 1 diabetic mice. The hypoglycemic effect of taurine was associated with the improvement of hepatic glucose metabolism, which includes upregulation of the hepatic glucose uptake via GLUT-2 and recovery of the hepatic glycogen content. Suppression of serum ketone body levels by taurine reflects improved hepatic glucose metabolism and energy production. Tissue MDA levels reduced by taurine suggest that the anti-oxidative action of taurine is also involved in the anti-diabetic effect of taurine in STZ-induced diabetic mice.

Author Contributions: S.M. conceived and designed the research. S.M., K.F. and N.T. performed the animal experiments and biochemical analysis. T.I. performed the microarray and pathway analyses. M.N. performed the histological study. S.M. wrote, reviewed, and edited the manuscript with support from T.I. and M.N. All authors have read and agreed to the published version of the manuscript.

Funding: This research received no external funding.

Institutional Review Board Statement: The study was conducted according to the guidelines of the Declaration of Helsinki and all experimental procedures were approved by the Animal Ethics Committee of Fukui Prefectural University (Approval Number 17-10).

Informed Consent Statement: Not applicable.

Data Availability Statement: The data presented in this study are available in the article.

Conflicts of Interest: The authors declare no conflict of interest.

Abbreviations

GK	Glucokinase
GLUT-2	Glucose transporter-2
G6Pase	Glucose-6-phosphatase
MDA	Malondialdehyde
8-OHdG	8-Hydroxy-2′-deoxyguanosine
PEPCK	Phosphoenolpyruvate carboxykinase
ROS	Reactive oxygen species
SREBP-1c	Sterol regulatory element-binding protein 1c
STZ	Streptozotocin

References

1. Huxtable, R.J. Physiological actions of taurine. *Physiol. Rev.* **1992**, *72*, 101–163. [CrossRef] [PubMed]
2. Lambert, I.H.; Kristensen, D.M.; Holm, J.B.; Mortensen, O.H. Physiological role of taurine—From organism to organelle. *Acta Physiol.* **2015**, *213*, 191–212. [CrossRef] [PubMed]
3. Sturman, J.A.; Rassin, D.K.; Gaull, G.E. Taurine in development. *Life Sci.* **1977**, *21*, 1–22. [CrossRef]
4. Albrecht, J.; Schousboe, A. Taurine interaction with neurotransmitter receptors in the CNS: An update. *Neurochem. Res.* **2005**, *30*, 1615–1621. [CrossRef]
5. Spriet, L.L.; Whitfield, J. Taurine and skeletal muscle function. *Curr. Opin. Clin. Nutr. Metab. Care* **2015**, *18*, 96–101. [CrossRef]
6. Lombardini, J.B. Taurine: Retinal function. *Brain Res. Rev.* **1991**, *16*, 151–169. [CrossRef]
7. Jong, C.J.; Sandal, P.; Schaffer, S.W. The role of taurine in mitochondria health: More than just an antioxidant. *Molecules* **2021**, *26*, 4913. [CrossRef]
8. Hardison, W.G. Hepatic taurine concentration and dietary taurine as regulators of bile acid conjugation with taurine. *Gastroenterology* **1978**, *75*, 71–75. [CrossRef]
9. Zulli, A. Taurine in cardiovascular disease. *Curr. Opin. Clin. Nutr. Metab. Care* **2011**, *14*, 57–60. [CrossRef]
10. Menzie, J.; Pan, C.; Prentice, H.; Wu, J.Y. Taurine and central nervous system disorders. *Amino Acids* **2014**, *46*, 31–46. [CrossRef]
11. Murakami, S. Role of taurine in the pathogenesis of obesity. *Mol. Nutr. Food Res.* **2015**, *59*, 1353–1363. [CrossRef]
12. Miyazaki, T.; Matsuzaki, Y. Taurine and liver diseases: A focus on the heterogeneous protective properties of taurine. *Amino Acids* **2014**, *46*, 101–110. [CrossRef]
13. Marcinkiewicz, J.; Kontny, E. Taurine and inflammatory diseases. *Amino Acids* **2014**, *46*, 7–20. [CrossRef]
14. Warskulat, U.; Heller-Stilb, B.; Oermann, E.; Zilles, K.; Haas, H.; Lang, F.; Häussinger, D. Phenotype of the taurine transporter knockout mouse. *Methods Enzymol.* **2007**, *428*, 439–458.
15. Warskulat, U.; Flögel, U.; Jacoby, C.; Hartwig, H.G.; Thewissen, M.; Merx, M.W.; Molojavyi, A.; Heller-Stilb, B.; Schrader, J.; Häussinger, D. Taurine transporter knockout depletes muscle taurine levels and results in severe skeletal muscle impairment but leaves cardiac function uncompromised. *FASEB J.* **2004**, *18*, 577–579. [CrossRef]
16. Ito, T.; Kimura, Y.; Uozumi, Y.; Takai, M.; Muraoka, S.; Matsuda, T.; Ueki, K.; Yoshiyama, M.; Ikawa, M.; Okabe, M.; et al. Taurine depletion caused by knocking out the taurine transporter gene leads to cardiomyopathy with cardiac atrophy. *J. Mol. Cell. Cardiol.* **2008**, *44*, 927–937. [CrossRef]
17. Ito, T.; Yoshikawa, N.; Inui, T.; Miyazaki, N.; Schaffer, S.W.; Azuma, J. Tissue depletion of taurine accelerates skeletal muscle senescence and leads to early death in mice. *PLoS ONE* **2014**, *9*, e107409. [CrossRef]
18. Franconi, F.; Loizzo, A.; Ghirlanda, G.; Seghieri, G. Taurine supplementation and diabetes mellitus. *Curr. Opin. Clin. Nutr. Metab. Care* **2006**, *9*, 32–36. [CrossRef]
19. Imae, M.; Asano, T.; Murakami, S. Potential role of taurine in the prevention of diabetes and metabolic syndrome. *Amino Acids* **2014**, *46*, 81–88. [CrossRef]
20. Tenner, T.E., Jr.; Zhang, X.J.; Lombardini, J.B. Hypoglycemic effects of taurine in the alloxan-treated rabbit, a model for type 1 diabetes. *Adv. Exp. Med. Biol.* **2003**, *526*, 97–104.
21. Trachtman, H.; Futterweit, S.; Maesaka, J.; Ma, C.; Valderrama, E.; Fuchs, A.; Tarectecan, A.A.; Rao, P.S.; Sturman, J.; Boles, T.H.; et al. Taurine ameliorates chronic streptozocin-induced diabetic nephropathy in rats. *Am. J. Physiol.* **1995**, *269*, F429–F438. [CrossRef]
22. Das, J.; Sil, P.C. Taurine ameliorates alloxan-induced diabetic renal injury, oxidative stress-related signaling pathways and apoptosis in rats. *Amino Acids* **2012**, *43*, 1509–1523. [CrossRef]
23. Nandhini, A.T.; Thirunavukkarasu, V.; Anuradha, C.V. Taurine modifies insulin signaling enzymes in the fructose-fed insulin resistant rats. *Diabetes Metab.* **2005**, *31*, 337–344. [CrossRef]

24. Nakaya, Y.; Minami, A.; Harada, N.; Sakamoto, S.; Niwa, Y.; Ohnaka, M. Taurine improves insulin sensitivity in the Otsuka Long-Evans Tokushima Fatty rat, a model of spontaneous type 2 diabetes. *Am. J. Clin. Nutr.* **2000**, *71*, 54–58. [CrossRef] [PubMed]
25. Park, E.J.; Bae, J.H.; Kim, S.Y.; Lim, J.G.; Baek, W.K.; Kwon, T.K.; Suh, S.I.; Park, J.W.; Lee, I.K.; Ashcroft, F.M.; et al. Inhibition of ATP-sensitive K+ channels by taurine through a benzamido-binding site on sulfonylurea receptor 1. *Biochem. Pharmacol.* **2004**, *67*, 1089–1096. [CrossRef] [PubMed]
26. Ribeiro, R.A.; Vanzela, E.C.; Oliveira, C.A.; Bonfleur, M.L.; Boschero, A.C.; Carneiro, E.M. Taurine supplementation: Involvement of cholinergic/phospholipase C and protein kinase A pathways in potentiation of insulin secretion and Ca^{2+} handling in mouse pancreatic islets. *Br. J. Nutr.* **2010**, *104*, 1148–1155. [CrossRef] [PubMed]
27. Carneiro, E.M.; Latorraca, M.Q.; Araujo, E.; Beltrá, M.; Oliveras, M.J.; Navarro, M.; Berná, G.; Bedoya, F.J.; Velloso, L.A.; Soria, B.; et al. Taurine supplementation modulates glucose homeostasis and islet function. *J. Nutr. Biochem.* **2009**, *20*, 503–511. [CrossRef] [PubMed]
28. Wu, N.; Lu, Y.; He, B.; Zhang, Y.; Lin, J.; Zhao, S.; Zhang, W.; Li, Y.; Han, P. Taurine prevents free fatty acid-induced hepatic insulin resistance in association with inhibiting JNK1 activation and improving insulin signaling in vivo. *Diabetes Res. Clin. Pract.* **2010**, *90*, 288–296. [CrossRef]
29. Kulakowski, E.C.; Maturo, J. Hypoglycemic properties of taurine: Not mediated by enhanced insulin release. *Biochem. Pharmacol.* **1984**, *33*, 2835–2838. [CrossRef]
30. Maturo, J.; Kulakowski, E.C. Taurine binding to the purified insulin receptor. *Biochem. Pharmacol.* **1988**, *37*, 3755–3760. [CrossRef]
31. Borck, P.C.; Vettorazzi, J.F.; Branco, R.C.S.; Batista, T.M.; Santos-Silva, J.C.; Nakanishi, V.Y.; Boschero, A.C.; Ribeiro, R.A.; Carneiro, E.M. Taurine supplementation induces long-term beneficial effects on glucose homeostasis in ob/ob mice. *Amino Acids* **2018**, *50*, 765–774. [CrossRef]
32. Di Leo, M.; Santini, S.A.; Silveri, N.G.; Giardina, B.; Franconi, F.; Ghirlanda, G. Long-term taurine supplementation reduces mortality rate in streptozotocin-induced diabetic rats. *Amino Acids* **2004**, *27*, 187–191. [CrossRef]
33. Das, J.; Vasan, V.; Sil, P.C. Taurine exerts hypoglycemic effect in alloxan-induced diabetic rats, improves insulin-mediated glucose transport signaling pathway in heart and ameliorates cardiac oxidative stress and apoptosis. *Toxicol. Appl. Pharmacol.* **2012**, *258*, 296–308. [CrossRef]
34. Barthel, A.; Schmoll, D. Novel concepts in insulin regulation of hepatic gluconeogenesis. *Am. J. Physiol. Endocrinol. Metab.* **2003**, *285*, E685–E692. [CrossRef]
35. Thorens, B. GLUT2, glucose sensing and glucose homeostasis. *Diabetologia* **2015**, *58*, 221–232. [CrossRef]
36. Rui, L. Energy metabolism in the liver. *Compr. Physiol.* **2014**, *4*, 177–197.
37. Kanikarla-Marie, P.; Jain, S.K. Hyperketonemia and ketosis increase the risk of complications in type 1 diabetes. *Free Radic. Biol. Med.* **2016**, *95*, 268–277. [CrossRef]
38. Tokunaga, H.; Yoneda, Y.; Kuriyama, K. Streptozotocin-induced elevation of pancreatic taurine content and suppressive effect of taurine on insulin secretion. *Eur. J. Pharmacol.* **1983**, *87*, 237–243. [CrossRef]
39. Lin, S.; Wu, G.; Zhao, D.; Han, J.; Yang, Q.; Feng, Y.; Liu, M.; Yang, J.; Hu, J. Taurine increases insulin expression in STZ-treated rat islet cells in vitro. *Adv. Exp. Med. Biol.* **2017**, *975*, 319–328.
40. Giacco, F.; Brownlee, M. Oxidative stress and diabetic complications. *Circ. Res.* **2010**, *107*, 1058–1070. [CrossRef]
41. Jha, J.C.; Banal, C.; Chow, B.S.; Cooper, M.E.; Jandeleit-Dahm, K. Diabetes and kidney disease: Role of oxidative stress. *Antioxid. Redox Signal.* **2016**, *25*, 657–684. [CrossRef]
42. Hansen, S.H. The role of taurine in diabetes and the development of diabetic complications. *Diabetes Metab. Res. Rev.* **2001**, *17*, 330–346. [CrossRef]
43. Ito, T.; Schaffer, S.W.; Azuma, J. The potential usefulness of taurine on diabetes mellitus and its complications. *Amino Acids* **2012**, *42*, 1529–1539. [CrossRef]
44. Pushpakiran, G.; Mahalakshmi, K.; Anuradha, C.V. Taurine restores ethanol-induced depletion of antioxidants and attenuates oxidative stress in rat tissues. *Amino Acids* **2004**, *27*, 91–96. [CrossRef]
45. Das, J.; Ghosh, J.; Manna, P.; Sil, P.C. Taurine protects acetaminophen-induced oxidative damage in mice kidney through APAP urinary excretion and CYP2E1 inactivation. *Toxicology* **2010**, *269*, 24–34. [CrossRef]
46. Li, S.; Wei, B.K.; Wang, J.; Dong, G.; Wang, X. Taurine supplementation ameliorates arsenic-induced hepatotoxicity and oxidative stress in mouse. *Adv. Exp. Med. Biol.* **2019**, *1155*, 463–470.
47. Murakami, S.; Ono, A.; Kawasaki, A.; Takenaga, T.; Ito, T. Taurine attenuates the development of hepatic steatosis through the inhibition of oxidative stress in a model of nonalcoholic fatty liver disease in vivo and in vitro. *Amino Acids* **2018**, *50*, 1279–1288. [CrossRef]
48. Sullivan, M.A.; Forbes, J.M. Glucose and glycogen in the diabetic kidney: Heroes or villains? *EBioMedicine* **2019**, *47*, 590–597. [CrossRef] [PubMed]
49. Nannipieri, M.; Lanfranchi, A.; Santerini, D.; Catalano, C.; Van de Werve, G.; Ferrannini, E. Influence of long-term diabetes on renal glycogen metabolism in the rat. *Nephron* **2001**, *87*, 50–57. [CrossRef] [PubMed]
50. Abdulrazaq, N.B.; Cho, M.M.; Win, N.N.; Zaman, R.; Rahman, M.T. Beneficial effects of ginger (Zingiber officinale) on carbohydrate metabolism in streptozotocin-induced diabetic rats. *Br. J. Nutr.* **2012**, *108*, 1194–1201. [CrossRef] [PubMed]
51. Luc, K.; Schramm-Luc, A.; Guzik, T.J.; Mikolajczyk, T.P. Oxidative stress and inflammatory markers in prediabetes and diabetes. *J. Physiol. Pharmacol.* **2019**, *70*, 809–824.

52. De Luca, G.; Calpona, P.R.; Caponetti, A.; Romano, G.; Di Benedetto, A.; Cucinotta, D.; Di Giorgio, R.M. Taurine and osmoregulation: Platelet taurine content, uptake, and release in type 2 diabetic patients. *Metabolism* **2001**, *50*, 60–64. [CrossRef]
53. Reibel, D.K.; Shaffer, J.E.; Kocsis, J.J.; Neely, J.R. Changes in taurine content in heart and other organs of diabetic rats. *J. Mol. Cell Cardiol.* **1979**, *11*, 827–830. [CrossRef]
54. Nguyen-Ngo, C.; Willcox, J.C.; Lappas, M. Anti-diabetic, anti-inflammatory, and anti-oxidant effects of naringenin in an in vitro human model and an in vivo murine model of gestational diabetes mellitus. *Mol. Nutr. Food Res.* **2019**, *63*, e1900224. [CrossRef]
55. Shi, G.J.; Li, Y.; Cao, Q.H.; Wu, H.X.; Tang, X.Y.; Gao, X.H.; Yu, J.Q.; Chen, Z.; Yang, Y. In vitro and in vivo evidence that quercetin protects against diabetes and its complications: A systematic review of the literature. *Biomed. Pharmacother.* **2019**, *109*, 1085–1099. [CrossRef]
56. Ito, T.; Yoshikawa, N.; Ito, H.; Schaffer, S.W. Impact of taurine depletion on glucose control and insulin secretion in mice. *J. Pharmacol. Sci.* **2015**, *129*, 59–64. [CrossRef]
57. Ma, N.; Aoki, E.; Semba, R. An immunohistochemical study of aspartate, glutamate, and taurine in rat kidney. *J. Histochem. Cytochem.* **1994**, *42*, 621–626. [CrossRef]

Article

Hepatic Transcriptome Analysis Provides New Insight into the Lipid-Reducing Effect of Dietary Taurine in High–Fat Fed Groupers (*Epinephelus coioides*)

Mingfan Chen [1,†], Fakai Bai [1,†], Tao Song [1], Xingjian Niu [1], Xuexi Wang [2], Kun Wang [1] and Jidan Ye [1,*]

[1] Xiamen Key Laboratory for Feed Quality Testing and Safety Evaluation, Fisheries College, Jimei University, Xiamen 361021, China; 202111908029@jmu.edu.cn (M.C.); 201911908010@jmu.edu.cn (F.B.); 202011908003@jmu.edu.cn (T.S.); niuqinjian@neau.edu.cn (X.N.); wangkun@jmu.edu.cn (K.W.)

[2] Key Laboratory of Marine Biotechnology of Fujian Province, College of Marine Sciences, Fujian Agriculture and Forestry University, Fuzhou 350002, China; w1508498684@sina.com or wangxuexi@fafu.edu.cn

* Correspondence: yjyjd@jmu.edu.cn; Tel.: +86-59-2618-1420

† These authors contributed equally to this work.

Abstract: A transcriptome analysis was conducted to provide the first detailed overview of dietary taurine intervention on liver lipid accumulation caused by high–fat in groupers. After an eight-week feeding, the fish fed 15% fat diet (High–fat diet) had higher liver lipid contents vs. fish fed 10% fat diet (Control diet). 15% fat diet with 1% taurine (Taurine diet) improved weight gain and feed utilization, and decreased hepatosomatic index and liver lipid contents vs. the High–fat diet. In the comparison of the Control vs. High–fat groups, a total of 160 differentially expressed genes (DEGs) were identified, of which up- and down-regulated genes were 72 and 88, respectively. There were 49 identified DEGs with 26 and 23 of up- and down-regulated in the comparison to High–fat vs. Taurine. Several key genes, such as cysteine dioxygenase (*CDO1*), ADP–ribosylation factor 1/2 (*ARF1_2*), sodium/potassium–transporting ATPase subunit alpha (*ATP1A*), carnitine/acylcarnitine translocase (*CACT*), and calcium/calmodulin–dependent protein kinase II (*CAMK*) were obtained by enrichment for the above DEGs. These genes were enriched in taurine and hypotaurine metabolism, bile secretion, insulin secretion, phospholipase D signaling pathway, and thermogenesis pathways, respectively. The present study will also provide a new insight into the nutritional physiological function of taurine in farmed fish.

Keywords: taurine; fat metabolism; liver fat accumulation; RNA–Seq; *Epinephelus coioides*

Citation: Chen, M.; Bai, F.; Song, T.; Niu, X.; Wang, X.; Wang, K.; Ye, J. Hepatic Transcriptome Analysis Provides New Insight into the Lipid-Reducing Effect of Dietary Taurine in High–Fat Fed Groupers (*Epinephelus coioides*). *Metabolites* **2022**, *12*, 670. https://doi.org/10.3390/metabo12070670

Academic Editors: Teruo Miyazaki, Takashi Ito, Alessia Baseggio Conrado and Shigeru Murakami

Received: 5 July 2022
Accepted: 19 July 2022
Published: 20 July 2022

1. Introduction

Taurine is a sulfur–containing amino acid which is the most abundant free amino acid in animals [1]. It is clear that dietary taurine administration can reduce peripheral cholesterol and visceral lipid accumulation of rats and humans by enhancing rate–limiting enzyme activity of cholesterol 7α–hydroxylase in the liver, promote the synthesis of cholic acid from cholesterol, and increase the excretion of fecal cholesterol [2–4]. Taurine is a conditionally essential amino acid for most cultured fish [5]. It plays a range of key roles in fish physiology, including functions in bile acid conjugation, immune regulation, osmoregulation, antioxidation, nervous system development, and regeneration [6,7]. Moreover, the lipid-reducing effect of dietary taurine also occurred in different tissues of many farmed fish [8–11].

At present, one of the particularly concerning issues is fatty liver syndrome caused by widespread use of high–fat feed in intensive aquaculture for the purpose of protein-sparing and feed utilization [12,13]. Despite these benefits, excessive fat intake does cause growth retardation [14,15] and other undesirable effects, such as visceral lipid accumulation and fatty liver [16–19], accompanied by apoptosis and declined immune function [19–21]. Therefore, fatty liver induced by high–fat feeding has become a typical chronic liver disease which is closely associated with nutritional metabolic syndrome in intensive fish farming. There is an urgent need to find a suitable way to solve the problem of fatty livers, which represents a threat to aquaculture production. In the light of the lipid–reducing effect of taurine mentioned above, dietary taurine administration may therefore be a promising way to attenuate the adverse effects caused by high–fat diets.

The orange–spotted grouper (*Epinephelus coioides*) has become an economically important mariculture carnivorous fish species in Southeast Asian countries, including China [22,23]. Previous studies showed that dietary taurine supplementation can attenuate the tissue lipid accumulation of groupers [11,24]. However, the underlying mechanism involved in the regulation of lipid metabolism still remains unclear. Recent studies have shown that dietary taurine supplementation in 15% fat feed could reduce lipid accumulation through reducing the contents of triglyceride molecules containing 18:2n–6 at the *sn–2* and *sn–3* positions [25] and through accelerating lipid absorption of taurine−conjugated bile acids and fatty acid β-oxidation, and inhibiting lipogenesis [26]. However, determining how dietary taurine modulates lipid metabolism of fish at the transcriptional level has not yet been investigated. In the present study, three experimental diets (10% fat, 15% fat and 15% fat with 1% taurine) were formulated to investigate the lipid–reducing effect of dietary taurine supplementation in regard to taurine-mediated changes of key genes.

2. Results

2.1. Growth Performance and Tissue Lipid Contents

The growth performance of groupers is shown in Table 1. When the dietary lipid level increased from 10% to 15%, weight gain, feed conversion ratio and hepatosomatic index of the Control and High–fat groups were not different ($p > 0.05$). However, when 15% lipid diet was added with 1% taurine, weight gain was significantly enhanced, whereas feed conversion ratio and hepatosomatic index were markedly reduced compared with High–fat group ($p < 0.05$).

Table 1. Effects of experimental diets on growth performance and the contents of liver and muscle fat of groupers (*E. coioides*).

Parameters	Diets (Fat Level/Taurine Level)		
	Control (10/0)	**High–Fat (15/0)**	**Taurine (15/1)**
Weight gain (%)	151.44 ± 9.86^{a}	157.38 ± 4.77^{a}	180.07 ± 3.45^{b}
Feed conversion ratio	0.99 ± 0.01^{b}	1.00 ± 0.01^{b}	0.91 ± 0.02^{a}
Hepatosomatic index (%)	2.79 ± 0.14^{b}	2.89 ± 0.06^{b}	1.56 ± 0.24^{a}
Liver lipid content (%)	9.57 ± 0.56^{a}	13.38 ± 0.47^{b}	9.05 ± 0.31^{a}
Muscle lipid content (%)	1.31 ± 0.72^{a}	1.95 ± 0.11^{b}	1.79 ± 0.01^{b}

Weight gain (%) = 100 × (final body weight − initial body weight)/initial body weight; Feed conversion ratio = feed intake/wet weight gain; Hepatosomatic index (%) = 100 × (liver weight/wet body weight); Data are presented as the means ± SEM (n = 3 tanks); Data was presented as the means ± SEM (n = 30 fish); Values in the same row with different lowercase letter superscripts indicate significant differences ($p < 0.05$); Statistical analysis was performed by one way ANOVA, followed by Student–Neuman–Keuls multiple comparison test.

The lipid contents in the liver and muscle samples in High–fat group were significantly higher ($p < 0.05$) than that in Control group. Fish of Taurine group had lower ($p < 0.05$) liver lipid contents vs those of High–fat group. However, there was no difference in muscle lipid content between High–fat and Taurine groups ($p > 0.05$).

2.2. Illumina Sequencing and De Novo Assembly

According to the sequencing results of the transcriptome assay, we obtained 399.25 million clean reads (Table S1), with Q20 at between 97.56% and 98.46%, Q30 at between 93.16% and 95.45%, and the GC content between 42.16% and 47.00%. From the assembly results, 99,634 unigenes were obtained with the N50 and N90 of unigenes being 653 and 243 bp respectively (Table 2). The length distribution of transcript and unigenes is shown in Figure S1. After short and low–quality sequences were excluded, 349,211 unigenes were identified and annotated by matching them against the five public databases (NR, SwissProt, PFAM, GO and KO), and 72,679 unigenes had at least one Blast hit against the public databases, yielding 72,125 annotated unigenes for NR (20.65%), 13,104 for SwissProt (3.75%), 22,814 for PFAM (6.53%), 6988 for GO (2.00%) and 5124 for KO (1.47%) (Table 3).

Table 2. Summary statistics of the de novo transcriptome assembly of liver samples of *E. coioides*.

Items	Min Length	Max Length	Mean Length	N50	N90
Unigene number	200	16,130	532	653	243

Min length, the minimum sequence length in a unigene set; Max length, the maximum sequence length in a unigene set; Mean length, average sequence length of a unigene set; N50/N90, the unigenes were calculated by ordering all sequences, and the length of unigenes was then collected one by one from the longest to the shortest until 50%/90% of the total length was attained.

Table 3. The number of annotated genes in different databases.

Type of Database Annotation	Number of Unigenes	Percentage (%)
NR	72,125	20.65
SwissProt	13,104	3.75
PFAM	22,814	6.53
GO	6988	2.00
KO	5124	1.47
Annotated in all databases	487	0.14
Annotated in at least one database	72,679	20.81
Total Unigenes	349,211	100

NR, NCBI non-redundant protein sequence database; SwissProt, protein sequence database; PFAM, a large collection of protein multiple sequence alignments and profile hidden Markov models; GO, gene ontology database; KO, KEGG orthology database for representation of gene/protein functional orthologs in molecular networks; KEGG, kyoto encyclopedia of genes and genome database.

2.3. Identification of DEGs

To identify DEGs in liver samples of grouper after feeding trial, three digital gene expression libraries from the Control, High–fat, and Taurine groups were constructed (Figure 1). There were 160 DEGs identified in the comparison between the High–fat and Control groups, of which 72 were upregulated and 88 were downregulated. In the comparison between the Taurine and High–fat groups, 49 DEGs were identified, of which 26 were upregulated and the rest were downregulated.

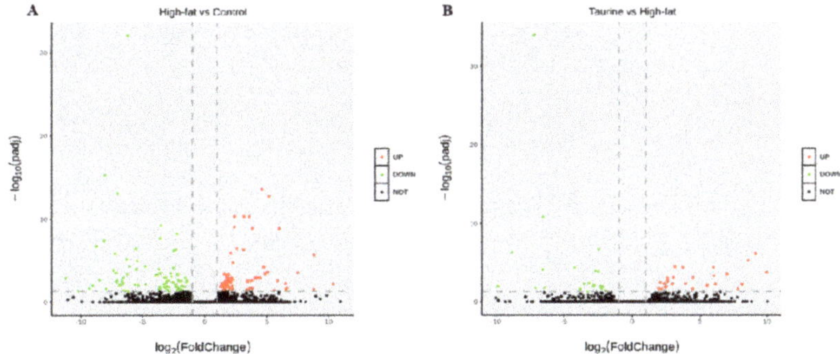

Figure 1. Comparison of DEGs among liver transcriptome of *E. coioides* of the Control and the High-fat groups, and of the High–fat and the Taurine groups. Control, control diet; High–fat, 15% fat diet; Taurine, 15% fat diet with 1% taurine. (**A**)—volcano plot of DEGs of High-fat vs. Control group; (**B**)—volcano plot of DEGs of Taurine vs. High–fat group.

2.4. Function Annotation and Analysis for Unigenes

Based upon gene ontology (GO) classification, 6988 (2.00%) unigenes were mapped and clustered into biological processes, cellular components and molecular function categories. In the classification of biological processes, the cellular process and metabolic process occur most frequently. In the cellular components, most unigenes were classified into a cell and cell part. With regard to molecular function, most unigenes were clustered into binding and catalytic activity categories. The GO annotation statistics of Unigenes is shown in Figure 2.

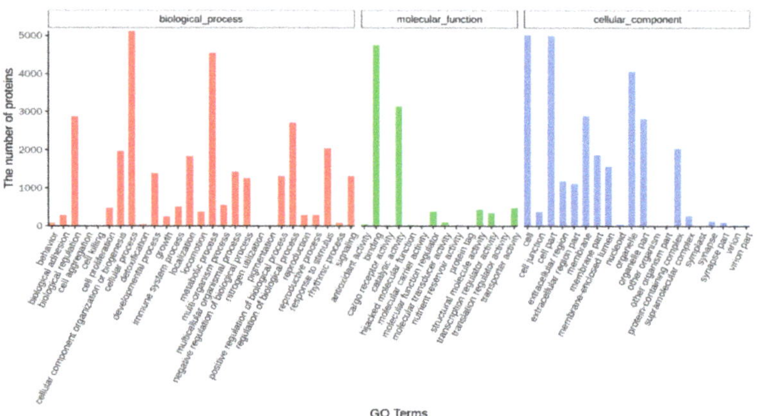

Figure 2. Classification of the unigenes and annotation of differentially expressed genes (DEGs) of liver samples of *E. coioides* based on enrichment analysis of gene ontology. All DEGs were enriched into three categorizations: biological processes, cellular components and molecular function categories.

Upon KEGG enrichment analysis, 160 differentially expressed genes (DEGs) in the liver samples were identified in the comparison with the High–fat vs. the Control groups, of which 72 genes were upregulated and 88 genes were downregulated. The typical pathways of DEGs for enrichment were complement and coagulation cascades, respectively (Figure 3). The comparison of Taurine and High–fat groups had 49 DEGs (26 genes were upregulated and the rest were downregulated) for enrichment in the liver. The insulin secretion was the most enriched pathway of DEGs (Figure 3).

2.5. Signaling Pathway Network Related to Lipid Metabolism

According to the above analysis results, the DEGs were annotated as key genes such as cysteine dioxygenase (*CDO1*), ADP–ribosylation factor 1/2 (*ARF1_2*), sodium/potassium–transporting ATPase subunit alpha (ATP1α) in the carnitine/acylcarnitine translocase (CACT) and calcium/calmodulin-dependent protein kinase II (CAMK). These key genes were high–fat induced lipid metabolism genes regulated by dietary taurine in groupers. Several signaling pathways of taurine–mediated lipid metabolism were clustered in Table S2 and mainly included taurine and hypotaurine metabolism, primary bile secretion, insulin secretion, phospholipase D signaling pathway, phosphatidylinositol signaling system, inositol phosphate metabolism and calcium signaling pathway. Furthermore, the signaling pathway network related to taurine–mediated lipid metabolism in the liver was constructed by KEGG functional enrichment pathway conjoint analysis and is presented in Figure 4.

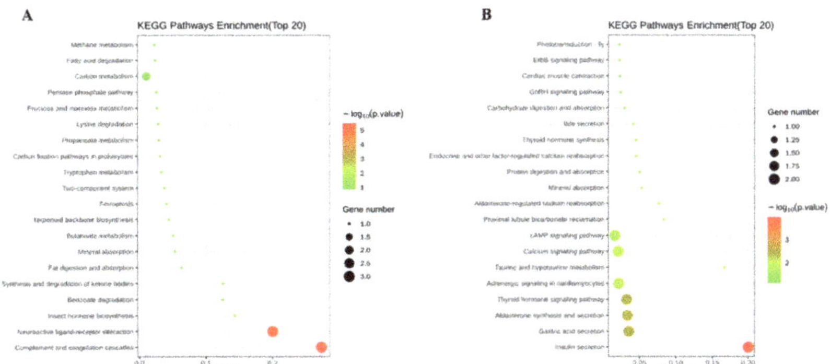

Figure 3. Bubble chart of significantly enriched KEGG pathways in the differentially expressed genes (DEGs) of liver samples of *E. coioides* between the Control and High–fat groups, and between High–fat and Taurine groups. (**A**)—pathway enrichment of DEGs of High–fat vs. Control groups; (**B**)—pathway enrichment of DEGs of High–fat vs. Taurine groups. The vertical axis represents the pathway categories, the horizontal axis shows the enrichment factor. The point size shows the number of DEGs enriched in the KEGG pathway. The point color shows different Q values as indicated on the right.

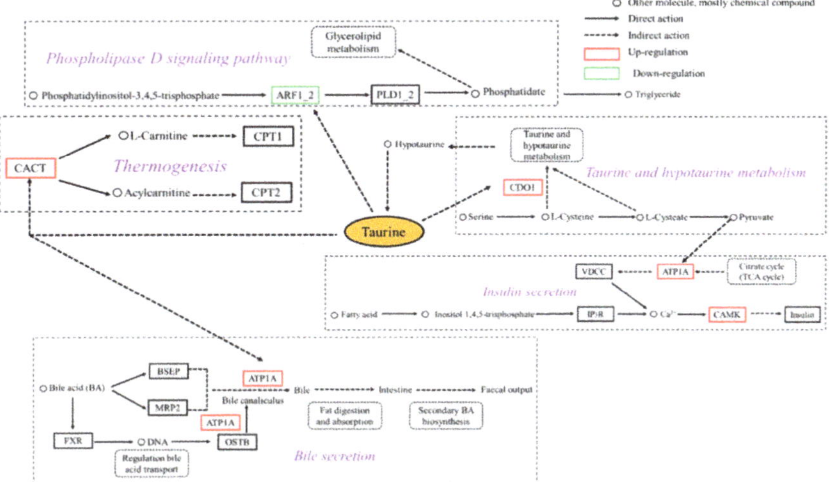

Figure 4. Signal pathway network diagram of taurine–mediated lipid metabolism in the liver of *E. coioides*. Abbreviations: AACT, carnitine/acylcarnitine translocase; ARF1_2, ADP–ribosylation factor 1/2; ATP1α, sodium/potassium–transporting ATPase subunit alpha; CAMK, calmodulin–dependent protein kinase II; CDO1, cysteine dioxygenase; GAPN, glyceraldehyde-3-phosphate dehydrogenase; GK, glucokinase; PDK1, pyruvate dehydrogenase kinase isozyme 2; PLCD, phosphatidylinositol phospholipase C.

2.6. qRT–PCR Validation of DEGs

To validate the reliability of RNA–Seq data, eight randomly selected DEG expression profiles in taurine and high fat comparison groups samples were examined by qRT–PCR. As shown in Figure 5, the fold–changes obtained by qRT–PCR were consistent with the values obtained by RNA–seq for the six selected genes (*ARF1_2, GK, ATP1*α*, CAMK, CDO1,* and *CACT*), suggesting our RNA-Seq data and the results based on RNA–Seq data analysis were reliable.

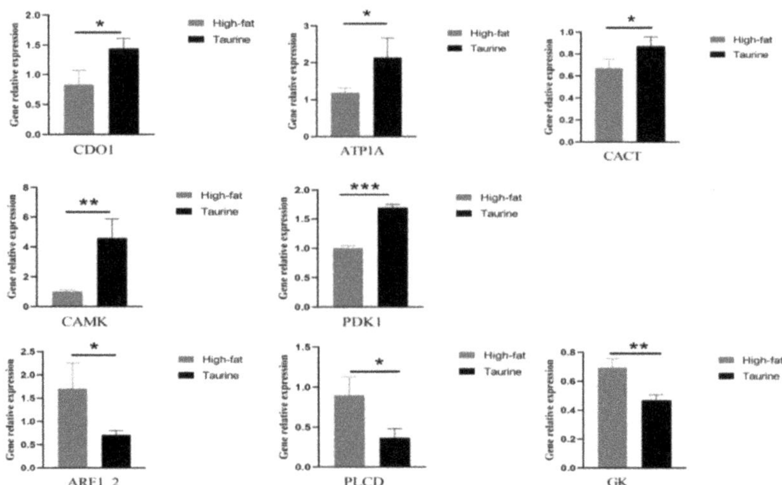

Figure 5. Validation of RNA–seq data using qRT–PCR in the liver of *E. coioides* in High–fat and Taurine groups. To validate the data from RNA–seq analysis, relative mRNA levels of eight selected differentially expressed genes (*ARF1_2, GK, ATP1α, CAMK, CDO1, CACT, PDK1,* and *PLCD*) from the liver samples of Control, High–fat and Taurine groups were examined by RT–qPCR. mRNA levels are presented as –fold changes when compared with the Control group after normalization against β-actin. The relative mRNA levels from the RNA–seq analysis were calculated as FPKM values. High–fat, 15% fat diet; Taurine, 15% fat diet with 1% taurine. Student's *t*–test was applied for comparison of High–fat vs. Taurine diets. Values are means ± SEM (n = 3). Asterisks (*, ** and ***) represent significant differences with $p < 0.05$, $p < 0.01$ and $p < 0.001$ respectively.

3. Discussion

The results of the present study showed that the growth and feed utilization of groupers did not differ when the dietary fat was increased from 10% to 15%, which indicates that the fish species has a certain degree of tolerance to a higher fat diet. Similar results were observed in previous studies of black sea bream (*Acanthopagrus schlegelii*) [27] and large yellow croaker (*Larimichthys crocea*) [28]. However, high–fat diets increased liver lipid contents vs. control diets in this study and previous studies on many farmed fishes [16–18,20,28–30], indicating that feeding high–fat diet led to an increase in liver lipid accumulation of fish.

Interestingly, taurine addition in high-fat diets not only promoted both fish growth and feed utilization in the study and previous studies on many farmed fishes [31–35], but also reduced liver lipid content and/or hepatosomatic index of farmed fishes [10,11,36–40]. All of these results indicate a significant effect of taurine on reducing liver lipid. In view of the lipid–reducing effect of taurine on fish above mentioned, the next objective of this study was to investigate how high–fat diet does affect liver lipid deposition and to explore the lipid-reducing effect of dietary taurine intervention in relation to taurine–mediated changes of gene expression at the transcriptome level.

Herein, we present the first transcriptomic analysis of taurine intervention on the liver lipid metabolism of high–fat fed groupers in an attempt to understand the regulatory mechanism of taurine on liver lipid deposition caused by high–fat feeding. The DEGs were then screened out after the unigenes were mapped to five public databases (NR, SwissProt, PFAM, GO and KO) by the NCBI, and had at least one Blast hit against the public databases. A total of 160 DEGs were identified in the comparison of High–fat and Control groups. In the comparison of High–fat and Taurine groups, a total of 49 DEGs were identified. GO annotation and KEGG pathway analysis of DEGs in the comparison of Taurine and High–fat groups showed that the DEGs were enriched in primary bile secretion, insulin secretion, phospholipase D signaling pathway and thermogenesis pathways. The discovery of these genes and signaling pathways should contribute to a better understanding of the molecular mechanism of regulation of taurine in fish lipid metabolism.

The cysteine sulfite dependent pathway is the main route of taurine biosynthesis in mammals, in which CDO1, as the rate–limiting enzyme, can produce hypotaurine through oxidation and decarboxylation, then hypotaurine is oxidized to form taurine [41]. Fish also have the capability for biosynthesis of taurine by CDO1 [6]. The activity of CDO1 in fish liver is generally higher than that in other tissues [5]. Dietary taurine addition was reported to upregulate the gene expression of *CDO1* in tissues of Atlantic bluefin tuna (*Thunnus thynnus*) [42], but did not affect its expression in the liver of tiger puffer (*Takifugu rubripes*) [43] and European bass (*Dicentrarchus labrax*) [8]. The dietary taurine promotes liver *CDO1* expression level in a dose–dependent manner [37]. The above results show that the ability of fish to synthesize taurine varies between fish species [37]. In this study, there was no difference in expression level of *CDO1* between Control and High–fat diets, but the Taurine diet resulted in an upregulation in gene expression of *CDO1* compared with the High–fat diet. The *CDO1* gene was enriched in the taurine and hypotaurine metabolic pathway. One possible reason for this is the fact that feeding high–fat diet may increase the demand for taurine of groupers. This indicates that taurine exerts the effect of reducing liver fat by affecting the synthesis and metabolism of taurine.

In the present study, dietary taurine supplementation downregulated the expression of ADP–ribosylation factor (ARF) gene *ARF1_2* compared with the High–fat group. The *ARF1_2* gene was enriched into phospholipase D signaling pathway. The *ARF1_2* gene is a member of the Ras superfamily, and participates in vesicular trafficking and regulating the activation of phospholipase D (PLD), which plays an important role in intracellular signal transduction and substance transport [44]. PLD catalyzes not only the hydrolysis of the phosphodiester bond of glycerophospholipids to generate phosphatidic acid (PA), but also a transphosphatidylation reaction to produce phosphatidylethanol [45]. The PLD activated by ARF1_2 hydrolyzed phosphatidylcholine to produce more precursor PA required for triglyceride synthesis [46,47]. Therefore, dietary taurine supplementation may reduce the synthesis of triglycerides in the liver of fish through downregulating *ARF1_2* gene expression.

As a signal molecule, bile acids are able to regulate their own enterohepatic circulation by affecting transcription of the genes critically involved in transport and metabolism [48]. The farnesoid X receptor (FXR) is activated by bile acids [49]. The activated FXR triggers the transcriptional synthesis of bile salt export pump (BSEP) [50]. Due to the high affinity of BSEP to bile acids, the bile acids in the form of complex bile acids and BSEP are transported from hepatocytes to the intestine with the help of the energy generated by ATP hydrolysis. BSEP is therefore considered to be the most important ATP dependent bile acid transporter in the liver [51]. The bile acid metabolism and transport are achieved through the chain reaction triggered by FXR and its downstream BSEP [50,52], so as to maintain the glycolipid metabolic homeostasis of fish [18,53]. Dietary taurine administration ameliorates Na^+/K^+ATPase impairment in the retina of diabetic rats [54] and promotes gill Na^+/K^+ATPase activity of rainbow trout (*Oncorhynchus mykiss*) [55]. In this study, after dietary taurine intervention in the High–fat group, *ATP1α* gene expression was upregulated and enriched into the bile secretion pathway. Therefore, dietary taurine administration accelerates bile acid transport and metabolism through upregulating *ATP1α* gene expression.

Lipid metabolic disorder is associated with abnormal energy metabolism, including gluconeogenesis, glycogenolysis, and the TCA cycle. GK is a rate–limiting enzyme required for glucose metabolism in the liver and the regulation of insulin secretion from islets [56]. Hepatic GK enzyme activity of glycolysis pathway is stimulated by glucose uptake [57], contributing an increase in liver glycogen content to maintain glucose homeostasis. In contrast, the downregulation of hepatic *GK* gene and protein expression was accompanied by reduced hepatic glycogen synthesis in type 2 diabetic rats [58]. The findings of the present study showed that dietary taurine supplementation could inhibit the glycolysis pathway as well as the biosynthesis of pyruvate, which is the final metabolites of glycolysis and the substrate for lipid synthesis de novo. Therefore, the lipid–reducing effect of taurine on fish can also be indirectly achieved by inhibiting the biosynthesis of substrate of lipogenesis de novo via down–regulating the gene expression in the glycolysis pathway.

The activation of CAMK II was involved in the hypertonicity–induced upregulation of human taurine transporter [59], enhancing insulin secretion [60]. In this study, the expression of gene *CAMK* (Ca^{2+}/Calmodulin kinase II) was up-regulated after dietary taurine intervention, and enriched into the insulin secretion signal pathway. Taurine can promote insulin secretion through the interaction between taurine and ATP sensitive K^+ (KATP) channels [61,62].

Insulin is an important endocrine hormone. It promotes free fatty acid and cholesterol uptake, reduces lipolysis, and increases lipogenesis in fish [63] through regulating several enzymes involved in lipogenesis and lipolysis, as well as transcription factors regulating the expression of such enzymes [64,65]. High–fat diets caused lipid metabolism disorder, insulin resistance, and liver steatosis in mammals [66], and liver steatosis was positively associated with insulin resistance in nonalcoholic fatty liver disease [67]. Dietary taurine administration can increase the insulin sensitivity of mammals [68,69]. Therefore, dietary taurine may promote insulin secretion through up regulating the expression of *CAMK* gene. On the one hand, it may inhibit insulin resistance caused by a high–fat diet. On the other hand, it is conducive to the decomposition of peripheral free fatty acids and cholesterol into the liver, thus reducing peripheral fat deposition. However, the relevant results here still need further study in this regard.

Liver lipid homeostasis regulation involves a complex interaction of triglycerides present in hepatocytes including fatty acid uptake, de novo lipogenesis, fatty acid β-oxidation, and fatty acid export [66,70,71]. Abnormal liver fat accumulation is often accompanied by excessive production of reactive oxygen free radicals and intermediates with lipotoxicity such as diacylglycerol as well as endoplasmic reticulum stress, resulting in lipid metabolism disorder [71,72]. Mitochondria are recognized as the main organelles of fatty acid β-oxidation. The fatty acid β-oxidation is regulated by gene *CACT* [73]. The transcriptional level of *CACT* gene was up–regulated by fish oil both in rats and fish compared with beef tallow or lard oil [74,75]. In this study, dietary taurine intervention resulted in an upregulation of the *CACT* gene expression. The *CACT* gene was enriched into the thermogenesis pathway. This indicates that up–regulated *CACT* gene expression caused by dietary taurine intervention promotes fatty acid transport to mitochondria and fatty acid β-oxidation, thereby reducing lipid deposition in the liver of groupers.

4. Materials and Methods

4.1. Experimental Diets

The optimal dietary levels of lipid and taurine were at about 10% and 1%, respectively for the growth of *E. coioides* [76,77]. In this experiment, therefore, three isonitrogenous (46% crude protein) experimental diets were prepared using casein and gelatin without taurine (food grade) and shrimp meal as protein source, fish oil, soy oil and soy lecithin as lipid source, namely 10% fat det (control diet), 15% fat diet (high fat diet) and 15% fat + 1% taurine diet (Taurine diet). The ingredients and proximate composition of the three experimental diets are shown in Table 4. The experimental diets were produced according to the method described in detail previously [24].

4.2. Growth Trial

Juvenile groupers purchased from a local fish farm (Zhangpu county, Fujian province, China) were transported to the Aquaculture Experimental Center of Jimei University. They were fed a commercial feed for two weeks of acclimatization in two tanks (1000 L/tank) before the start of the experiment. A total of 270 fish with a similar size (an initial wet body weight of 10.5 ± 0.1 g) were randomly distributed into 9 fiberglass tanks at a stocking density of 30 fish per tank (300 L/tank), within a recirculating water aquaculture system connected with a circulation pump, biological filters, and an automatic temperature control device. The fish in triplicate tanks of each group were fed one of the three experimental diets to apparent satiation twice daily (8:30 and 18:30), across a feeding period of 8 weeks. Excess feed was collected 30 min after each meal, dried in a ventilated oven at 65 °C, and weighed for the calculation of feed intake. During the feeding period, water temperature was maintained at about 28.0 °C, dissolved oxygen level ranged between 6.1 ± 0.2 mg/L, and the ammonia nitrogen level was less than 0.2 mg/L.

Table 4. Ingredients and proximate composition of experimental diets (on an as–fed basis, %) [1].

Ingredients	Diets (Fat Level/Taurine Level)		
	Control (10/0)	**High-Fat (15/0)**	**Taurine (15/1)**
Animal by–product (casein:gelatin = 4:1)	50	50	50
Shrimp meal	4	4	4
Corn starch	25	25	25
Oil (fish:soy oil = 1:1)	6	10	10
Soy lecithin	4	4	4
Premix	0.8	0.8	0.8
$Ca(H_2PO_4)_2$	2	2	2
Microcrystalline cellulose	7.2	3.2	2.2
Sodium alginate	1	1	1
Taurine	0	0	1
Nutrient level (analyzed values)			
Dry matter	91.18	90.24	90.38
Crude protein	46.55	46.87	46.56
Crude lipid	10.44	14.79	14.89
Taurine	0.04	0.04	0.98

[1] All feed ingredients are provided by Jiakang Feed Co., Ltd. (Xiamen, China).

4.3. Sample Collection

At the end of the 8-week feeding trial, the fish in each tank were caught and anesthetized with a dose of 100 mg/L solution of MS-222 (tricaine methane sulphonate, Sigma-Aldrich Shanghai Trading Co. Ltd., Shanghai, China), followed by fish count and batch–weighing, and recorded on a wet weight basis to determine percent weight gain and feed conversion ratio. Five fish were randomly caught from each tank and individually weighed. The liver was then removed after abdominal dissection, and weighed for the determination of hepatosomatic index. The liver samples were pooled by tank, frozen immediately in liquid nitrogen and stored at −20 °C for the analysis of fat content. For gene expression analysis, another three fish in each tank were randomly sampled and anesthetized. After drawing blood, liver samples were aseptically removed and quickly frozen with liquid nitrogen and stored at −80 °C.

4.4. Composition Analysis

Proximate composition of ingredients, diets, and tissue samples were determined according to standard methods [78]. Dry matter was determined by drying the samples in an oven at 105 °C to a constant weight. Crude protein was determined by the Kjeldahl method (N × 6.25) using Kjeltec TM 8400 Auto Sample Systems (Foss Teacher AB). The crude fat content was determined by the Soxtec extraction method by using Soxtec Avanti 2050 (Foss Teacher AB). Ash was measured in the residues of samples burned in a muffle furnace at 550 °C for 8 h.

Total fat of muscle and liver samples were extracted by homogenization in chloroform/methanol (2:1, v/v) solution and determined gravimetrically after drying a 5 mL aliquot under nitrogen [79]. For taurine determination [80], feed samples were hydrolysed in nitrogen–flushed glass vials using 6 mol/L HCl at 116 °C for 22 h, followed by centrifugation (1500 g, 4 °C, 15 min), and the supernatant was collected and analyzed using an automatic AA analyzer (Hitachi L8900, Tokyo, Japan).

4.5. RNA Extraction and cDNA Library Construction

Total RNA was extracted from the liver sample using TRIzol® reagent (Magen) according to the manufacturer's instructions. The A260/A280 absorbance ratios of RNA in the liver samples were detected to test the RNA purity by using NanoDrop ND–2000 spectrophotometer (Thermo Scientific, Waltham, MA, USA). The RIN value of sample RNA was detected to test the RNA quality of the samples by using Agilent Bioanalyzer 4150 (Agilent Technologies, Santa Clara, CA, USA). The cDNA library was constructed using the Illumina HiseqTM 2000 system (Illumina, San Diego, CA, USA) by APTBIO Co., Ltd. (Shanghai, China). Raw reads were filtered to remove low-quality sequences using the program written by APTBIO Co., Ltd.

4.6. Sequence Data Processing and Analysis

RNA–Seq de novo assembly was carried out by using Trinity software (http://trinityrnaseq.github.io/) (accessed on 27 July 2020). The longest transcripts were regarded as unigenes after removing repetitive assemblies. BLAST software (http://blast.ncbi.nlm.nih.gov/Blast.cgi/) (accessed on 27 July 2020) was used to align unigene sequences with the Non-Redundant Protein Sequence Database (NR, https://ftp.ncbi.nlm.nih.gov/blast/db/FASTA/) (accessed on 27 July 2020), Protein Families Database (Pfam, http://pfam.xfam.org/) (accessed on 27 July 2020), SwissProt protein Database (SwissProt, https://www.expasy.org/) (accessed on 27 July 2020), Kyoto Encyclopedia of Gene and Genomes Database (KEGG, http://www.kegg.jp/) (accessed on 27 July 2020), and Gene Ontology Database (GO, http://geneontology.org) (accessed on 27 July 2020). RESM software (http://deweylab.github.io/RSEM/) (accessed on 27 July 2020) was used to accurately map the sequencing reads to reference genomes. The expression level of each gene was calculated from the fragment per kilobase of exon model per million mapped read (FPKM) values [81].

4.7. Identification and Enrichment Analysis of Differentially Expressed Genes

DESeq2 software (https://bioconductor.org/packages/release/bioc/html/DESeq2.html) (accessed on 27 July 2020) was used to determine differentially expressed genes (DEGs), and unigenes with p-value < 0.05 and |log2 foldchange| > 1 were defined as DEGs. In the process of DEGs analysis, the recognized and effective Benjamin Hochberg method was used to correct the significance p-value obtained from the original hypothesis test. The corrected p-value, FDR (false discovery rate), is then used as the key index of DEG screening to reduce the false positive caused by independent statistical hypothesis test on the expression value of a large number of genes. The number of up- and down-regulated DEGs in the liver under different dietary treatments was obtained. Using Blast2go software (https://www.blast2go.com/) (accessed on 27 July 2020), the gene ontology (GO) annotation information of all DEGs was obtained, the GO function of DEGs was classified, and the molecular function, cell composition and biological process of target genes were described [82]. The pathway enrichment analysis was performed using online service tool KAAS (KEGG automatic annotation server). Fisher's precision probability test was used to calculate the significance of enrichment of each gene in the pathway, so as to determine the corresponding significant signal transduction and metabolic pathways. The enrichment results of DEGs are displayed by KEGG enrichment scatter diagram.

4.8. Quantitative Real–Time PCR Analysis

The relative expression levels of eight selected DEGs were verified by quantitative reverse transcription PCR (qRT–PCR), so as to validate the gene expression data obtained by RNA–Seq. The eight pairs of primers were designed with Primer v. 5.0, with the β-actin gene of the fish used as the internal reference for the qPCR analysis (Table S3) and sent to jinweizhi Biotechnology Co., Ltd. (Suzhou, China) for synthesis. The qRT–PCR reactions were performed using ChamQ Universal SYBR qPCR Master Mix (Vazyme, Nanjing, China) on a Real–Time PCR Detection System (ABI 7500, Applied Biosystems, Waltham, MA, USA). The PCR cycling conditions were as follows: 95 °C for 30 s, followed by 40 cycles of 95 °C for 10 s and 60 °C for 30 s, and then cycles at 95 °C for 15 s, 60 °C for 60 s and 95 °C for 15 s. To check reproducibility, the qRT–PCR reaction for each sample was performed in four biological replicates. The relative expression of genes was calculated using the $2^{-\Delta\Delta Ct}$ method [83].

4.9. Statistical Analysis

Data are presented as mean and standard errors of the mean. Data were subjected to *t*-test and one-way ANOVA and Student–Neuman–Keuls multiple comparison tests in SPSS Statistics 22.0 (SPSS, Michigan Avenue, Chicago, IL, USA). $p < 0.05$ was considered statistically significant.

5. Conclusions

In summary, the results of the present study show that feeding 15% fat diets did not result in alterations in growth and feed utilization, but increased liver fat accumulation of groupers vs those subject to 10% fat diets. However, 1% taurine addition in a 15% fat diet not only improved its growth performance, but also reduced liver fat deposition. Liver transcriptome analysis showed that 49 DEGs were identified in the comparison of High–fat and Taurine groups, of which the expression of *CDO1*, *ATP1α*, *CAMK*, and *CACT* genes was significantly up-regulated and *ARF1_2* gene expression was significantly down–regulated. The key genes were involved in the taurine and hypotaurine metabolic pathway, bile secretion, insulin secretion, thermogenic pathway, and phospholipase D signaling pathway. As a result, the effect of dietary taurine on reducing liver fat accumulation of the fish species may be achieved by enhancing the synthesis of endogenous taurine in the liver, accelerating bile acid transport and promoting insulin secretion and fatty acid β-oxidation efficiency. The next step is to investigate the roles and functions of the key genes mentioned above in the fat metabolism of fish in response to dietary taurine.

Supplementary Materials: The following supporting information can be downloaded at: https://www.mdpi.com/article/10.3390/metabo12070670/s1. Figure S1: The sequence length of transcript and unigenes; Table S1: Alignment statistics of the RNA–Seq analysis of the nine liver samples of *E. coioides* between Control and High–fat groups, and between High–fat and Taurine groups; Table S2: The DEGs with significantly changed KEGG pathways related to lipid metabolism in the liver of *E. coioides* in the comparison of High–fat and Taurine groups; Table S3: Primers sequences of lipid metabolism related genes used for real–time PCR for *E. coioides*.

Author Contributions: Conceptualization, J.Y. and F.B.; methodology, M.C. and F.B.; software, M.C.; validation, F.B., T.S. and X.N.; formal analysis, X.W. and X.N.; investigation, M.C. and F.B.; data curation, M.C.; writing—original draft preparation, M.C.; writing—review and editing, J.Y; visualization, F.B.; supervision, J.Y.; project administration, K.W.; funding acquisition, J.Y. All authors have read and agreed to the published version of the manuscript.

Funding: This research was funded by the National Natural Science Foundation of China, grant numbers 31772861 and 31372546, and the Science and Technology Project of Fujian Province, China, grant number 2020N0012.

Institutional Review Board Statement: The animal study protocol was approved by the Ethics Committee of Jimei University, Xiamen, China (protocol code 2011–58 and approved by 20 December 2011).

Informed Consent Statement: Not applicable.

Data Availability Statement: The data that support the findings of this study are available from the corresponding author upon reasonable request; further inquiries can be directed to NCBI Database at https://www.ncbi.nlm.nih.gov/bioproject/PRJNA853943 (accessed on 29 June 2022).

Conflicts of Interest: The authors declare no conflict of interest.

References

1. Lambert, I.H.; Kristensen, D.M.; Holm, J.B.; Mortensen, O.H. Physiological role of taurine—From organism to organelle. *Acta Physiol.* **2015**, *213*, 191–212. [CrossRef]
2. Chen, W.; Guo, J.; Chang, P. The effect of taurine on cholesterol metabolism. *Mol. Nutr. Food Res.* **2012**, *56*, 681–690. [CrossRef] [PubMed]
3. Fukuda, N.; Yoshitama, A.; Sugita, S.; Fujita, M.; Murakami, S. Dietary taurine reduces hepatic secretion of cholesteryl ester and enhances fatty acid oxidation in rats fed a high-cholesterol diet. *J. Nutr. Sci. Vitaminol.* **2011**, *57*, 144–149. [CrossRef] [PubMed]
4. Brons, C.; Spohr, C.; Storgaard, H.; Dyerberg, J.; Vaag, A. Effect of taurine treatment on insulin secretion and action, and on serum lipid levels in overweight men with a genetic predisposition for type II diabetes mellitus. *Eur. J. Clin. Nutr.* **2004**, *58*, 1239–1247. [CrossRef] [PubMed]
5. Wang, X.; He, G.; Mai, K.; Xu, W.; Zhou, H. Differential regulation of taurine biosynthesis in rainbow trout and Japanese flounder. *Sci. Rep.* **2016**, *6*, 21231. [CrossRef] [PubMed]
6. Salze, G.P.; Davis, D.A. Taurine: A critical nutrient for future fish feeds. *Aquaculture* **2015**, *437*, 215–229. [CrossRef]
7. Xu, H.; Zhang, Q.; Kim, S.K.; Liao, Z.; Wei, Y.; Sun, B.; Jia, L.; Chi, S.; Liang, M. Dietary taurine stimulates the hepatic biosynthesis of both bile acids and cholesterol in the marine teleost, tiger puffer (*Takifugu rubripes*). *Br. J. Nutr.* **2020**, *123*, 1345–1356. [CrossRef]
8. Martins, N.; Diógenes, A.F.; Magalhães, R.; Matas, I.; Oliva-Teles, A.; Peres, H. Dietary taurine supplementation affects lipid metabolism and improves the oxidative status of European seabass (*Dicentrarchus labrax*) juveniles. *Aquaculture* **2021**, *531*, 735820. [CrossRef]
9. Li, M.; Lai, H.; Li, Q.; Gong, S.; Wang, R. Effects of dietary taurine on growth, immunity and hyperammonemia in juvenile yellow catfish *Pelteobagrus fulvidraco* fed all-plant protein diets. *Aquaculture* **2016**, *450*, 349–355. [CrossRef]
10. Garcia-Organista, A.A.; Mata-Sotres, J.A.; Viana, M.T.; Rombenso, A.N. The effects of high dietary methionine and taurine are not equal in terms of growth and lipid metabolism of juvenile California yellowtail (*Seriola dorsalis*). *Aquaculture* **2019**, *512*, 734304. [CrossRef]
11. Koven, W.; Peduel, A.; Gada, M.; Nixon, O.; Ucko, M. Taurine improves the performance of white grouper juveniles (*Epinephelus aeneus*) fed a reduced fish meal diet. *Aquaculture* **2016**, *460*, 8–14. [CrossRef]
12. Du, Z.Y.; Liu, Y.J.; Tian, L.X.; Wang, J.T.; Wang, Y.; Liang, G.Y. Effect of dietary lipid level on growth, feed utilization and body composition by juvenile grass carp (*Ctenopharyngodon idella*). *Aquac. Nutr.* **2005**, *11*, 139–146. [CrossRef]
13. Meng, Y.; Tian, H.; Hu, X.; Han, B.; Li, X.; Cangzhong, L.; Ma, R.; Kubilay, A. Effects of dietary lipid levels on the lipid deposition and metabolism of subadult triploid rainbow trout (*Oncorhynchus mykiss*). *Aquac. Nutr.* **2022**, *2022*, 6924835. [CrossRef]
14. Dai, Y.; Cao, X.; Zhang, D.; Li, X.; Liu, W.; Jiang, G. Chronic inflammation is a key to inducing liver injury in blunt snout bream (*Megalobrama amblycephala*) fed with high-fat diet. *Dev. Comp. Immunol.* **2019**, *97*, 28–37. [CrossRef]
15. Fan, Z.; Li, J.; Zhang, Y.; Wu, D.; Zheng, X.; Wang, C.; Wang, L. Excessive dietary lipid affecting growth performance, feed utilization, lipid deposition, and hepatopancreas lipometabolism of large-sized common carp (*Cyprinus carpio*). *Front. Nutr.* **2021**, *8*, 694426. [CrossRef]
16. Fei, S.; Xia, Y.; Chen, Z.; Liu, C.; Liu, H.; Han, D.; Jin, J.; Yang, Y.; Zhu, X.; Xie, S. A high-fat diet alters lipid accumulation and oxidative stress and reduces the disease resistance of overwintering hybrid yellow catfish (*Pelteobagrus fulvidraco* ♀× *P. vachelli* ♂). *Aquac. Rep.* **2022**, *23*, 101043. [CrossRef]
17. Guo, J.; Zhou, Y.; Zhao, H.; Chen, W.; Chen, Y.; Lin, S. Effect of dietary lipid level on growth, lipid metabolism and oxidative status of largemouth bass, *Micropterus salmoides*. *Aquaculture* **2019**, *506*, 394–400. [CrossRef]
18. Yin, P.; Xie, S.; Zhuang, Z.; He, X.; Tang, X.; Tian, L.; Liu, Y.; Niu, J. Dietary supplementation of bile acid attenuate adverse effects of high-fat diet on growth performance, antioxidant ability, lipid accumulation and intestinal health in juvenile largemouth bass (*Micropterus salmoides*). *Aquaculture* **2021**, *531*, 735864. [CrossRef]
19. Jia, R.; Cao, L.P.; Du, J.L.; He, Q.; Gu, Z.Y.; Jeney, G.; Xu, P.; Yin, G.J. Effects of high-fat diet on antioxidative status, apoptosis and inflammation in liver of tilapia (*Oreochromis niloticus*) via Nrf2, TLRs and JNK pathways. *Fish Shellfish Immun.* **2020**, *104*, 391–401. [CrossRef]
20. Zhou, Y.L.; Guo, J.L.; Tang, R.J.; Ma, H.J.; Chen, Y.J.; Lin, S.M. High dietary lipid level alters the growth, hepatic metabolism enzyme, and anti-oxidative capacity in juvenile largemouth bass *Micropterus salmoides*. *Fish Physiol. Biochem.* **2020**, *46*, 125–134. [CrossRef]
21. Cao, X.; Dai, Y.; Liu, M.; Yuan, X.; Wang, C.; Huang, Y.; Liu, W.; Jiang, G. High-fat diet induces aberrant hepatic lipid secretion in blunt snout bream by activating endoplasmic reticulum stress-associated IRE1/XBP1 pathway. *BBA Mol. Cell Biol. Lipids* **2019**, *1864*, 213–223. [CrossRef] [PubMed]
22. Feng, H.; Yi, K.; Qian, X.; Niu, X.; Sun, Y.; Ye, J. Growth and metabolic responses of juvenile grouper (*Epinephelus coioides*) to dietary methionine/cystine ratio at constant sulfur amino acid levels. *Aquaculture* **2020**, *518*, 734869. [CrossRef]
23. Shapawi, R.; Abdullah, F.C.; Senoo, S.; Mustafa, S. Nutrition, growth and resilience of tiger grouper (*Epinephelus fuscoguttatus*) × giant grouper (*Epinephelus lanceolatus*) hybrid—A review. *Rev. Aquac.* **2019**, *11*, 1285–1296. [CrossRef]
24. Bai, F.; Niu, X.; Wang, X.; Ye, J. Growth performance, biochemical composition and expression of lipid metabolism related genes in groupers (*Epinephelus coioides*) are altered by dietary taurine. *Aquac. Nutr.* **2021**, *27*, 2690–2702. [CrossRef]
25. Bai, F.; Wang, X.; Niu, X.; Shen, G.; Ye, J. Lipidomic profiling reveals the reducing lipid accumulation effect of dietary taurine in groupers (*Epinephelus coioides*). *Front. Mol. Biosci.* **2021**, *8*, 814318. [CrossRef]

26. Wang, X.; Bai, F.; Niu, X.; Sun, Y.; Ye, J. The lipid-lowering effect of dietary taurine in orange-spotted groupers (*Epinephelus coioides*) involves both bile acids and lipid metabolism. *Front. Mar. Sci.* **2022**, *9*, 859428. [CrossRef]
27. Jin, M.; Zhu, T.; Tocher, D.R.; Luo, J.; Shen, Y.; Li, X.; Pan, T.; Yuan, Y.; Betancor, M.B.; Jiao, L.; et al. Dietary fenofibrate attenuated high-fat-diet-induced lipid accumulation and inflammation response partly through regulation of pparalpha and sirt1 in juvenile black seabream (*Acanthopagrus schlegelii*). *Dev. Comp. Immunol.* **2020**, *109*, 103691. [CrossRef]
28. Ding, T.; Xu, N.; Liu, Y.; Du, J.; Xiang, X.; Xu, D.; Liu, Q.; Yin, Z.; Li, J.; Mai, K.; et al. Effect of dietary bile acid (BA) on the growth performance, body composition, antioxidant responses and expression of lipid metabolism-related genes of juvenile large yellow croaker (*Larimichthys crocea*) fed high-lipid diets. *Aquaculture* **2020**, *518*, 734768. [CrossRef]
29. Wang, J.; Liu, Y.; Tian, L.; Mai, K.; Du, Z.; Wang, Y.; Yang, H. Effect of dietary lipid level on growth performance, lipid deposition, hepatic lipogenesis in juvenile cobia (*Rachycentron canadum*). *Aquaculture* **2005**, *249*, 439–447. [CrossRef]
30. Han, T.; Li, X.; Wang, J.; Hu, S.; Jiang, Y.; Zhong, X. Effect of dietary lipid level on growth, feed utilization and body composition of juvenile giant croaker *Nibea japonica*. *Aquaculture* **2014**, *434*, 145–150. [CrossRef]
31. Matsunari, H.; Takeuchi, T.; Takahashi, M.; Mushiake, K. Effect of dietary taurine supplementation on growth performance of yellowtail juveniles *Seriola quinqueradiata*. *Fish. Sci.* **2005**, *71*, 1131–1135. [CrossRef]
32. Matsunari, H.; Furuita, H.; Yamamoto, T.; Kim, S.; Sakakura, Y.; Takeuchi, T. Effect of dietary taurine and cystine on growth performance of juvenile red sea bream *Pagrus major*. *Aquaculture* **2008**, *274*, 142–147. [CrossRef]
33. Lim, S.; Oh, D.; Khosravi, S.; Cha, J.; Park, S.; Kim, K.; Lee, K. Taurine is an essential nutrient for juvenile parrot fish *Oplegnathus fasciatus*. *Aquaculture* **2013**, *414–415*, 274–279. [CrossRef]
34. Lunger, A.N.; Mclean, E.; Gaylord, T.G.; Kuhn, D.; Craig, S.R. Taurine supplementation to alternative dietary proteins used in fish meal replacement enhances growth of juvenile cobia (*Rachycentron canadum*). *Aquaculture* **2007**, *271*, 401–410. [CrossRef]
35. Martins, N.; Estevão-Rodrigues, T.; Diógenes, A.F.; Diaz-Rosales, P.; Oliva-Teles, A.; Peres, H. Taurine requirement for growth and nitrogen accretion of European sea bass (*Dicentrarchus labrax*, L.) juveniles. *Aquaculture* **2018**, *494*, 19–25. [CrossRef]
36. Yun, B.; Ai, Q.; Mai, K.; Xu, W.; Qi, G.; Luo, Y. Synergistic effects of dietary cholesterol and taurine on growth performance and cholesterol metabolism in juvenile turbot (*Scophthalmus maximus* L.) fed high plant protein diets. *Aquaculture* **2012**, *324*, 85–91. [CrossRef]
37. Candebat, C.L.; Booth, M.; Codabaccus, M.B.; Pirozzi, I. Dietary methionine spares the requirement for taurine in juvenile yellowtail kingfish (*Seriola lalandi*). *Aquaculture* **2020**, *522*, 735090. [CrossRef]
38. De Moura, L.B.; Diógenes, A.F.; Campelo, D.A.V.; Almeida, F.L.A.D.; Pousão-Ferreira, P.M.; Furuya, W.M.; Oliva-Teles, A.; Peres, H. Taurine and methionine supplementation as a nutritional strategy for growth promotion of meagre (*Argyrosomus regius*) fed high plant protein diets. *Aquaculture* **2018**, *497*, 389–395. [CrossRef]
39. Hu, Y.; Yang, G.; Li, Z.; Hu, Y.; Zhong, L.; Zhou, Q.; Peng, M. Effect of dietary taurine supplementation on growth, digestive enzyme, immunity and resistant to dry stress of rice field eel (*Monopterus albus*) fed low fish meal diets. *Aquac. Res.* **2018**, *49*, 2108–2118. [CrossRef]
40. Dehghani, R.; Oujifard, A.; Mozanzadeh, M.T.; Morshedi, V.; Bagheri, D. Effects of dietary taurine on growth performance, antioxidant status, digestive enzymes activities and skin mucosal immune responses in yellowfin seabream, *Acanthopagrus latus*. *Aquaculture* **2020**, *517*, 734795. [CrossRef]
41. Jurkowska, H.; Roman, H.B.; Hirschberger, L.L.; Sasakura, K.; Nagano, T.; Hanaoka, K.; Krijt, J.; Stipanuk, M.H. Primary hepatocytes from mice lacking cysteine dioxygenase show increased cysteine concentrations and higher rates of metabolism of cysteine to hydrogen sulfide and thiosulfate. *Amino Acids* **2014**, *46*, 1353–1365. [CrossRef] [PubMed]
42. Betancor, M.B.; Laurent, G.R.; Ortega, A.; de la Gándara, F.; Tocher, D.R.; Mourente, G. Taurine metabolism and effects of inclusion levels in rotifer (*Brachionus rotundiformis*, Tschugunoff, 1921) on Atlantic bluefin tuna (*Thunnus thynnus*, L.) larvae. *Aquaculture* **2019**, *510*, 353–363. [CrossRef]
43. Wei, Y.; Zhang, Q.; Xu, H.; Liang, M. Taurine requirement and metabolism response of tiger puffer *Takifugu rubripes* to graded taurine supplementation. *Aquaculture* **2020**, *524*, 735237. [CrossRef]
44. Wang, L.; Su, H.; Wang, C.L.; Mu, C.H. Structure and functional mechanism of the ADP-ribosylation factor. *Chin. J. Cell Biol.* **2007**, *29*, 675–681. (In Chinese)
45. Mcdermott, M.; Wakelam, M.J.; Morris, A.J. Phospholipase D. *Biochem. Cell Biol.* **2004**, *82*, 225–253. [CrossRef]
46. Kalra, H.; Drummen, G.P.C.; Mathivanan, S. Focus on extracellular vesicles: Introducing the next small big thing. *Int. J. Mol. Sci.* **2016**, *17*, 170. [CrossRef]
47. Martinez-Sena, T.; Soluyanova, P.; Guzman, C.; Valdivielso, J.M.; Castell, J.V.; Jover, R. The vitamin D receptor regulates glycerolipid and phospholipid metabolism in human hepatocytes. *Biomolecules* **2020**, *10*, 493. [CrossRef]
48. Kullak-Ublick, G.A.; Beuers, U.; Paumgartner, G. Hepatobiliary transport. *J. Hepatol.* **2000**, *32*, 3–18. [CrossRef]
49. Deuschle, U.; Birkel, M.; Hambruch, E.; Hornberger, M.; Kinzel, O.; Perovic-Ottstadt, S.; Schulz, A.; Hahn, U.; Burnet, M.; Kremoser, C. The nuclear bile acid receptor FXR controls the liver derived tumor suppressor histidine-rich glycoprotein. *Int. J. Cancer* **2015**, *136*, 2693–2704. [CrossRef]
50. Chen, Y.; Song, X.; Valanejad, L.; Vasilenko, A.; More, V.; Qiu, X.; Chen, W.; Lai, Y.; Slitt, A.; Stoner, M.; et al. Bile salt export pump is dysregulated with altered farnesoid X receptor isoform expression in patients with hepatocellular carcinoma. *Hepatology* **2013**, *57*, 1530–1541. [CrossRef]

51. Xiong, X.L.; Ding, Y.; Chen, Z.L.; Wang, Y.; Liu, P.; Qin, H.; Zhou, L.S.; Zhang, L.L.; Huang, J.; Zhao, L. Emodin rescues intrahepatic cholestasis via stimulating FXR/BSEP pathway in promoting the canalicular export of accumulated bile. *Front. Pharmacol.* **2019**, *10*, 522. [CrossRef] [PubMed]
52. Kubitz, R.; Dröge, C.; Stindt, J.; Weissenberger, K.; Häussinger, D. The bile salt export pump (BSEP) in health and disease. *Clin. Res. Hepatol. Gas.* **2012**, *36*, 536–553. [CrossRef] [PubMed]
53. Liao, Z.; Sun, B.; Zhang, Q.; Jia, L.; Wei, Y.; Liang, M.; Xu, H. Dietary bile acids regulate the hepatic lipid homeostasis in tiger puffer fed normal or high-lipid diets. *Aquaculture* **2020**, *519*, 734935. [CrossRef]
54. Di Leo, M.A.; Santini, S.A.; Cercone, S.; Lepore, D.; Gentiloni, S.N.; Caputo, S.; Greco, A.V.; Giardina, B.; Franconi, F.; Ghirlanda, G. Chronic taurine supplementation ameliorates oxidative stress and Na^+ K^+ ATPase impairment in the retina of diabetic rats. *Amino Acids* **2002**, *23*, 401–406. [CrossRef]
55. Huang, M.; Yang, X.; Zhou, Y.; Ge, J.; Davis, D.A.; Dong, Y.; Gao, Q.; Dong, S. Growth, serum biochemical parameters, salinity tolerance and antioxidant enzyme activity of rainbow trout (*Oncorhynchus mykiss*) in response to dietary taurine levels. *Mar. Life Sci. Technol.* **2021**, *3*, 449–462. [CrossRef]
56. Haeusler, R.A.; Camastra, S.; Astiarraga, B.; Nannipieri, M.; Anselmino, M.; Ferrannini, E. Decreased expression of hepatic glucokinase in type 2 diabetes. *Mol. Metab.* **2015**, *4*, 222–226. [CrossRef]
57. Caseras, A.; Meton, I.; Vives, C.; Egea, M.; Fernandez, F.; Baanante, I.V. Nutritional regulation of glucose-6-phosphatase gene expression in liver of the gilthead sea bream (*Sparus aurata*). *Br. J. Nutr.* **2002**, *88*, 607–614. [CrossRef]
58. Li, L.; Yang, J.; Liu, B.; Zou, Y.; Sun, M.; Li, Z.; Yang, R.; Xu, X.; Zou, L.; Li, G.; et al. P2Y12 shRNA normalizes inflammatory dysfunctional hepatic glucokinase activity in type 2 diabetic rats. *Biomed. Pharmacother.* **2020**, *132*, 110803. [CrossRef]
59. Satsu, H.; Manabe, M.; Shimizu, M. Activation of Ca^{2+}/calmodulin-dependent protein kinase II is involved in hyperosmotic induction of the human taurine transporter. *FEBS Lett.* **2004**, *569*, 123–128. [CrossRef]
60. Lee, S.J.; Kim, H.E.; Choi, S.E.; Shin, H.C.; Kwag, W.J.; Lee, B.K.; Cho, K.W.; Kang, Y. Involvement of Ca^{2+}/calmodulin kinase II (CaMK II) in genistein-induced potentiation of leucine/glutamine-stimulated insulin secretion. *Mol. Cells* **2009**, *28*, 167–174. [CrossRef]
61. Vettorazzi, J.F.; Ribeiro, R.A.; Santos-Silva, J.C.; Borck, P.C.; Batista, T.M.; Nardelli, T.R.; Boschero, A.C.; Carneiro, E.M. Taurine supplementation increases K (ATP) channel protein content, improving Ca^{2+} handling and insulin secretion in islets from malnourished mice fed on a high-fat diet. *Amino Acids* **2014**, *46*, 2123–2136. [CrossRef] [PubMed]
62. Ribeiro, R.A.; Bonfleur, M.L.; Amaral, A.G.; Vanzela, E.C.; Rocco, S.A.; Boschero, A.C.; Carneiro, E.M. Taurine supplementation enhances nutrient-induced insulin secretion in pancreatic mice islets. *Diabetes Metab. Res. Rev.* **2009**, *25*, 370–379. [CrossRef] [PubMed]
63. Caruso, M.A.; Sheridan, M.A. New insights into the signaling system and function of insulin in fish. *Gen. Comp. Endocr.* **2011**, *173*, 227–247. [CrossRef] [PubMed]
64. Polakof, S.; Médale, F.; Skiba-Cassy, S.; Corraze, G.; Panserat, S. Molecular regulation of lipid metabolism in liver and muscle of rainbow trout subjected to acute and chronic insulin treatments. *Domest. Anim. Endocrinol.* **2010**, *39*, 26–33. [CrossRef]
65. Pan, Y.; Luo, Z.; Wu, K.; Zhang, L.; Xu, Y.; Chen, Q. Cloning, mRNA expression and transcriptional regulation of five retinoid X receptor subtypes in yellow catfish *Pelteobagrus fulvidraco* by insulin. *Gen. Comp. Endocr.* **2016**, *225*, 133–141. [CrossRef] [PubMed]
66. Lian, C.Y.; Zhai, Z.Z.; Li, Z.F.; Wang, L. High fat diet-triggered non-alcoholic fatty liver disease: A review of proposed mechanisms. *Chem. Biol. Interact.* **2020**, *330*, 109199. [CrossRef]
67. Sakurai, M.; Takamura, T.; Ota, T.; Ando, H.; Akahori, H.; Kaji, K.; Sasaki, M.; Nakanuma, Y.; Miura, K.; Kaneko, S. Liver steatosis, but not fibrosis, is associated with insulin resistance in nonalcoholic fatty liver disease. *J. Gastroenterol.* **2007**, *42*, 312–317. [CrossRef]
68. Batista, T.M.; Ribeiro, R.A.; Da, S.P.; Camargo, R.L.; Lollo, P.C.; Boschero, A.C.; Carneiro, E.M. Taurine supplementation improves liver glucose control in normal protein and malnourished mice fed a high-fat diet. *Mol. Nutr. Food Res.* **2013**, *57*, 423–434. [CrossRef]
69. Nakaya, Y.; Minami, A.; Harada, N.; Sakamoto, S.; Niwa, Y.; Ohnaka, M. Taurine improves insulin sensitivity in the Otsuka Long-Evans Tokushima Fatty rat, a model of spontaneous type 2 diabetes. *Am. J. Clin. Nutr.* **2000**, *71*, 54–58. [CrossRef]
70. Musso, G.; Gambino, R.; Cassader, M. Recent insights into hepatic lipid metabolism in non-alcoholic fatty liver disease (NAFLD). *Prog. Lipid Res.* **2009**, *48*, 1–26. [CrossRef]
71. Ipsen, D.H.; Lykkesfeldt, J.; Tveden-Nyborg, P. Molecular mechanisms of hepatic lipid accumulation in non-alcoholic fatty liver disease. *Cell. Mol. Life Sci.* **2018**, *75*, 3313–3327. [CrossRef] [PubMed]
72. Gentile, C.L.; Nivala, A.M.; Gonzales, J.C.; Pfaffenbach, K.T.; Wang, D.; Wei, Y.; Jiang, H.; Orlicky, D.J.; Petersen, D.R.; Pagliassotti, M.J.; et al. Experimental evidence for therapeutic potential of taurine in the treatment of nonalcoholic fatty liver disease. *Am. J. Physiol. Regul. Integr. Comp. Physiol.* **2011**, *301*, R1710–R1722. [CrossRef] [PubMed]
73. Giudetti, A.M.; Stanca, E.; Siculella, L.; Gnoni, G.V.; Damiano, F. Nutritional and hormonal regulation of citrate and carnitine/acylcarnitine transporters: Two mitochondrial carriers involved in fatty acid metabolism. *Int. J. Mol. Sci.* **2016**, *17*, 817. [CrossRef]
74. Tian, J.; Lu, R.; Ji, H.; Sun, J.; Li, C.; Liu, P.; Lei, C.; Chen, L.; Du, Z. Comparative analysis of the hepatopancreas transcriptome of grass carp (*Ctenopharyngodon idellus*) fed with lard oil and fish oil diets. *Gene* **2015**, *565*, 192–200. [CrossRef]

75. Priore, P.; Stanca, E.; Gnoni, G.V.; Siculella, L. Dietary fat types differently modulate the activity and expression of mitochondrial carnitine/acylcarnitine translocase in rat liver. *Biochim. Biophys. Acta* **2012**, *1821*, 1341–1349. [CrossRef] [PubMed]
76. Luo, Z.; Liu, Y.; Mai, K.; Tian, L.; Liu, D.; Tan, X.; Lin, H. Effect of dietary lipid level on growth performance, feed utilization and body composition of grouper *Epinephelus coioides* juveniles fed isonitrogenous diets in floating netcages. *Aquac. Int.* **2005**, *13*, 257–269. [CrossRef]
77. Zhou, M.W.; Wang, H.W.; Ye, J.D. Responses to growth performance, body composition and tissue free amino acid contents of grouper (*Epinephelus coioides*) to dietary taurine content. *Chin. J. Anim. Nutr.* **2015**, *27*, 785–794. (In Chinese)
78. Horwitz, W. *Official Methods of Analysis of AOAC International*, 16th ed.; AOAC International: Rockville, MD,USA, 1995.
79. Folch, J.; Lees, M.; Sloane Stanley, G.H. A simple method for the isolation and purification of total lipides from animal tissues. *J. Biol. Chem.* **1957**, *226*, 497–509. [CrossRef]
80. Aragão, C.; Colen, R.; Ferreira, S.; Pinto, W.; Conceição, L.E.C.; Dias, J. Microencapsulation of taurine in *Senegalese sole* diets improves its metabolic availability. *Aquaculture* **2014**, *431*, 53–58. [CrossRef]
81. Trapnell, C.; Williams, B.A.; Pertea, G.; Mortazavi, A.; Kwan, G.; van Baren, M.J.; Salzberg, S.L.; Wold, B.J.; Pachter, L. Transcript assembly and quantification by RNA-Seq reveals unannotated transcripts and isoform switching during cell differentiation. *Nat. Biotechnol.* **2010**, *28*, 511–515. [CrossRef]
82. Ye, J.; Fang, L.; Zheng, H.; Zhang, Y.; Chen, J.; Zhang, Z.; Wang, J.; Li, S.; Li, R.; Bolund, L.; et al. WEGO: A web tool for plotting GO annotations. *Nucleic Acids Res.* **2006**, *34*, W293–W297. [CrossRef] [PubMed]
83. Livak, K.J.; Schmittgen, T.D. Analysis of relative gene expression data using real-time quantitative PCR and the $2^{-\Delta\Delta CT}$ Method. *Methods* **2001**, *25*, 402–408. [CrossRef] [PubMed]

MDPI

Article

Signaling Pathway of Taurine-Induced Upregulation of TXNIP

Hideo Satsu [1,*], Yusuke Gondo [2], Hana Shimanaka [2], Masato Imae [3], Shigeru Murakami [4], Kenji Watari [1], Shunichi Wakabayashi [5], Sung-Joon Park [5], Kenta Nakai [5] and Makoto Shimizu [6]

1 Department of Biotechnology, Faculty of Engineering, Maebashi Institute of Technology, Gunma 371-0816, Japan; m1281045watari@gmail.com
2 Department of Applied Biological Chemistry, Graduate School of Agricultural and Life Sciences, The University of Tokyo, Tokyo 113-8657, Japan; glic-gws@hotmail.co.jp (Y.G.); xxx_.87@icloud.com (H.S.)
3 Research & Development Headquarters Self-Medication, Taisho Pharmaceutical Co., Ltd., Tokyo 170-8633, Japan; m-imae@taisho.co.jp
4 Department of Bioscience and Biotechnology, Fukui Prefectural University, Fukui 910-1195, Japan; murakami@fpu.ac.jp
5 Human Genome Center, The Institute of Medical Science, The University of Tokyo, Tokyo 108-8639, Japan; shunichi.wakabayashi@gmail.com (S.W.); sjpark@ims.u-tokyo.ac.jp (S.-J.P.); knakai@ims.u-tokyo.ac.jp (K.N.)
6 Research Center for Agricultural and Life Sciences, Tokyo University of Agriculture, Tokyo 156-8502, Japan; ms205346@nodai.ac.jp
* Correspondence: satsu@maebashi-it.ac.jp; Tel.: +81-27-265-7374

Abstract: Taurine, a sulfur-containing β-amino acid, is present at high concentrations in mammalian tissues and plays an important role in several essential biological processes. However, the genetic mechanisms involved in these physiological processes associated with taurine remain unclear. In this study, we investigated the regulatory mechanism underlying the taurine-induced transcriptional enhancement of the thioredoxin-interacting protein (TXNIP). The results showed that taurine significantly increased the luciferase activity of the human TXNIP promoter. Further, deletion analysis of the TXNIP promoter showed that taurine induced luciferase activity only in the TXNIP promoter region (+200 to +218). Furthermore, by employing a bioinformatic analysis using the TRANSFAC database, we focused on Tst-1 and Ets-1 as candidates involved in taurine-induced transcription and found that the mutation in the Ets-1 sequence did not enhance transcriptional activity by taurine. Additionally, chromatin immunoprecipitation assays indicated that the binding of Ets-1 to the TXNIP promoter region was enhanced by taurine. Taurine also increased the levels of phosphorylated Ets-1, indicating activation of Ets-1 pathway by taurine. Moreover, an ERK cascade inhibitor significantly suppressed the taurine-induced increase in TXNIP mRNA levels and transcriptional enhancement of TXNIP. These results suggest that taurine enhances TXNIP expression by activating transcription factor Ets-1 via the ERK cascade.

Keywords: taurine; TXNIP; Ets-1; ERK

Citation: Satsu, H.; Gondo, Y.; Shimanaka, H.; Imae, M.; Murakami, S.; Watari, K.; Wakabayashi, S.; Park, S.-J.; Nakai, K.; Shimizu, M. Signaling Pathway of Taurine-Induced Upregulation of TXNIP. *Metabolites* **2022**, *12*, 636. https://doi.org/10.3390/metabo12070636

Academic Editor: Cholsoon Jang

Received: 21 May 2022
Accepted: 8 July 2022
Published: 11 July 2022

Publisher's Note: MDPI stays neutral with regard to jurisdictional claims in published maps and institutional affiliations.

1. Introduction

Taurine (2-aminoethanesulfonic acid) is a free β-amino acid abundant in several tissues, such as muscles, the heart, brain, and eyes. Previous research indicates that taurine performs various functions, including as an antioxidant, osmoregulator, and bile acid conjugate, as well as in detoxification [1,2]. Further, taurine is reported to be essential for the development of fetuses and newborns [3,4].

Taurine is obtained directly from diet and is also synthesized endogenously from cysteine, which in turn is formed from methionine. Dietary taurine is absorbed by intestinal epithelial cells via the taurine transporter (TAUT, SLC6A6). We previously reported that TAUT is regulated by various extracellular conditions, such as adaptive regulation and hyperosmolarity, as well as by tumor necrosis factor-α (TNF-α) [5–7]. Further, considering that the intestinal TAUT was regulated by inflammatory cytokines [7], we previously investigated the relationship between taurine and inflammation and reported that taurine reduced colitis symptoms in a mouse model of DSS-induced colitis [8]. Furthermore, using DNA microarrays, we comprehensively analyzed the effects of taurine on overall gene expression in intestinal epithelial cells and found that taurine markedly enhanced the mRNA expression and transcriptional activity of the thioredoxin-interacting protein (TXNIP) [9].

TXNIP suppresses its activity by interacting with thioredoxin [10,11], and various physiological functions have been reported for TXNIP [11]. It is reported that TXNIP knock-out mice have significantly reduced adaptability to energy deprivation [12–16] because they develop metabolic abnormalities, such as reduced efficiency of fatty acid metabolism [15], development of hyperlipidemia [13,17], and hypoglycemic states [14,16]. Furthermore, TXNIP contributes to glycogenesis in the liver [18], and the relationship between TXNIP mutations and the development of diabetes and hypertension has been suggested [17,19]. TXNIP has also been shown to regulate cardiac hypertrophy [20]. Furthermore, immune enhancement by moderately strong expression of TXNIP is also considered [11]. Physiological functions in the intestinal tract have been reported by Takahashi et al. [21] that TXNIP mRNA expression in the site of inflammation in patients with ulcerative colitis, is decreased, which is thought to contribute to the development of ulcerative colitis. Additionally, a previous study has shown that taurine regulates the functions of human intestinal Caco-2 cells via TXNIP induction [22]. Thus, taurine regulates the mRNA expression of TXNIP; however, the mechanisms underlying its regulation remain unclear.

In the present study, we analyzed the transcription factors and signaling molecules involved in the taurine-induced increase in the transcriptional activity of TXNIP.

2. Results

2.1. Effect of Taurine on Luciferase Activity Involving the TXNIP Promoter Region

Using reporter analysis, we previously reported that taurine increased the promoter activity of TXNIP [9]. A reporter vector containing the promoter region of TXNIP (−1299/+256) was used. Therefore, we examined the effect of taurine on promoter activity in various regions of the TXNIP promoter, including (−1299/+256), (−109/+256), and (−39/+256), with truncation of the 5′-flanking region. As shown in Figure 1A, taurine increased the promoter activity of all three reporter vectors, suggesting that the promoter region (−39/+256) is essential for the taurine-induced increase in TXNIP promoter activity.

We also constructed a reporter vector containing the (−39/+256), (−39/+142), and (−39/+65) sequences of the promoter region of TXNIP. Taurine markedly increased reporter activity in the presence of a reporter vector containing the promoter region between (−39/+256), but did not increase activity in the case of (−39/+142) and (−39/+65) (Figure 1B). This result suggests that a taurine response element is in the TXNIP promoter (+143/+256).

Based on the findings that the taurine response element is contained in the TXNIP promoter region between +143 and +256 (Figure 1), we next constructed a reporter vector containing the promoter region of (+122/+256) and examined the effect of taurine on the promoter activity of TXNIP (+122/+256). A reporter vector containing the promoter region (−1299/+142) was used as a negative control. The promoter activity of TXNIP (+122/+256) was significantly increased by taurine, whereas that of TXNIP (−1299/+142) did not change (Figure 2). This result confirmed that the taurine response element exists in the TXNIP promoter (+122/+256).

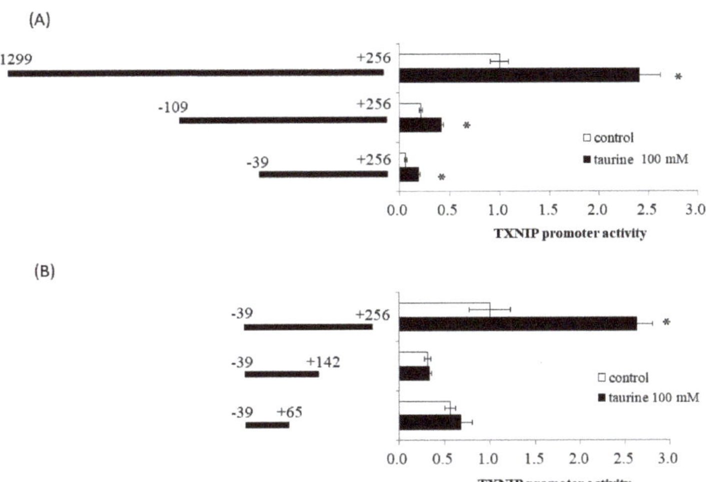

Figure 1. Effect of taurine on the transcriptional activity of the TXNIP promoter between −1299/+256. Caco-2 cells were transfected with a reporter vector containing partial promoter regions of TXNIP and then replaced with a medium containing 100 mM of taurine. After 24 h, the cells were subjected to a luciferase assay, as described in Materials and Methods Section 4.4. (**A**) TXNIP promoter partial sequences (−1299/+256, −109/+256, −39/+256). Results are expressed as relative values with the control value of (−1299/+256) as 1. (**B**) TXNIP promoter partial sequences (−29/+256, −29/+142, −39/+65). Results are expressed as relative values with the control value of −39/+256 as 1. Each value represents the mean ± S.E. (n = 3), * $p < 0.05$, vs. the control value (Student's t-test).

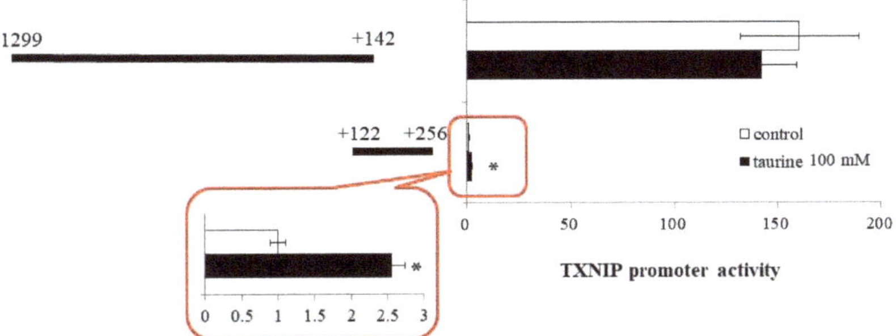

Figure 2. Effect of taurine on the transcriptional activity of the TXNIP promoter containing −1299/+142 and +122/+256. Caco-2 cells were transfected with a reporter vector containing partial promoter region of TXNIP (−1299/+142, +122/+256) and then replaced with a medium containing 100 mM of taurine. After 24 h, the cells were subjected to luciferase assay. The results are expressed as relative values, with a control value of +122/+256 as 1. Each value represents the mean ± S.E. (n = 3), * $p < 0.05$, vs. the control value (Student's t-test).

Then, the TXNIP promoter region (+122/+256) was divided into three regions (+122/+178, +162/+218, and +211/+256) and luciferase vectors each containing one region were constructed. Taurine significantly increased the promoter activity of TXNIP (+162/+218) (Figure 3), suggesting that the taurine response element is contained in the TXNIP promoter region (+162/+218).

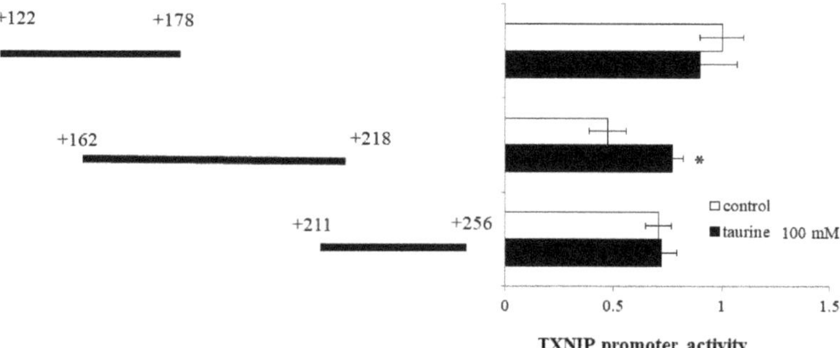

Figure 3. Effect of taurine on the transcriptional activity of the TXNIP promoter between +122/+256. Caco-2 cells were transfected with a reporter vector containing partial promoter regions of TXNIP (+122/+178, +162/+218, +211/+256) and then replaced with a medium containing 100 mM of taurine. After 24 h, the cells were subjected to luciferase assay. The results are expressed as relative values, with a control value of +122/+178 as 1. Each value represents the mean ± S.E. (n = 3), * $p < 0.05$, vs. the control value (Student's t-test).

Further, the TXNIP promoter region (+162/+218) was divided into three regions (+174/+191, +187/+204, and +200/+218), and a luciferase vector containing each region was constructed. Taurine significantly increased the promoter activity of TXNIP (+200/+218) (Figure 4), suggesting that the taurine response element is contained in the TXNIP promoter region (+200/+218). Taurine also increased the promoter activity of TXNIP (+200/+218) in a dose-dependent manner (Figure 5), confirming that a taurine response element exists in the promoter region (+200/+218).

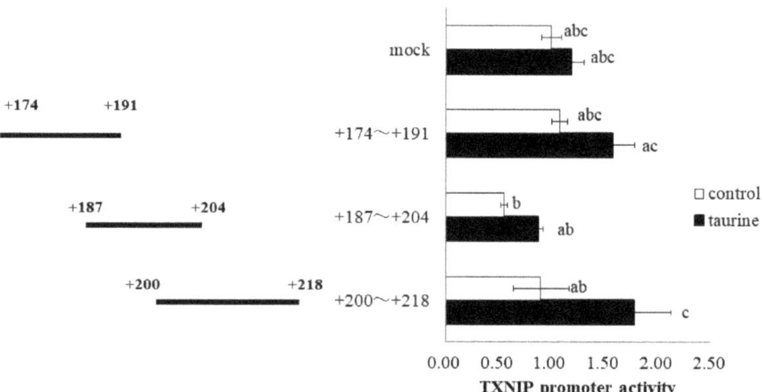

Figure 4. Effect of taurine on the transcriptional activity of the TXNIP promoter between +174/+218. Caco-2 cells were transfected with a reporter vector containing partial promoter regions of TXNIP (+174/+191, +187/+204, +200/+218) and then replaced with a medium containing 100 mM of taurine. After 24 h, the cells were subjected to a luciferase assay. The results are expressed as relative values, with the control value as 1. Each value is the mean ± S.E. (n = 3) and the [abc] values indicated by different characters are significantly different from each other, and the [abc] values indicated by the same characters are not significantly different (Tukey's test; $p < 0.05$).

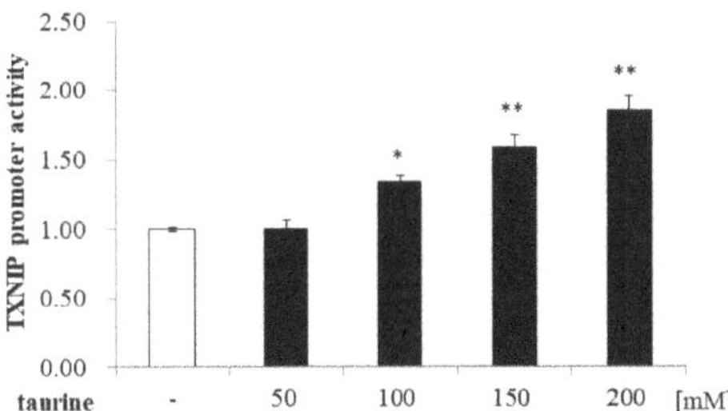

Figure 5. Dose-dependence of taurine-induced TXNIP promoter activity (+200/+218). Caco-2 cells were transfected with a reporter vector containing the partial promoter region of TXNIP (+200/+218) and then replaced with a medium containing 0–200 mM of taurine. After 24 h, the cells were subjected to luciferase assay. The results are expressed as relative values, with a control value of 1. Each value represents the mean ± S.E. ($n = 3$), * $p < 0.05$, ** $p < 0.01$, vs. the control value (Dunnett's test).

The promoter region of human TXNIP (+162/+218) was analyzed using a bioinformatics approach involving the TRANSFAC database. At the same time, deletion analysis of human TXNIP promoter was also performed. Then, the result of Figure 4 was obtained and the taurine-response region was narrowed down to +200/+218. Therefore, we especially analyzed the TXNIP promoter region (+200/+218) using TRANSFAC database and focused on Tst-1 and Ets-1 as candidates involved in taurine-induced transcription. We then constructed the human TXNIP (+211/+256) promoter vector with site-directed mutagenesis of putative Tst-1 and Ets-1 response elements, respectively (Figure 6A). The results showed that the mutation of the Tst-1 response element had no effect on the increase in TXNIP promoter activity induced by taurine (Figure 6B), but the mutation of the Ets-1 response element abolished the induction of TXNIP promoter activity by taurine (Figure 6C). This result strongly suggests that Ets-1 is involved in taurine-induced upregulation of TXNIP promoter activity.

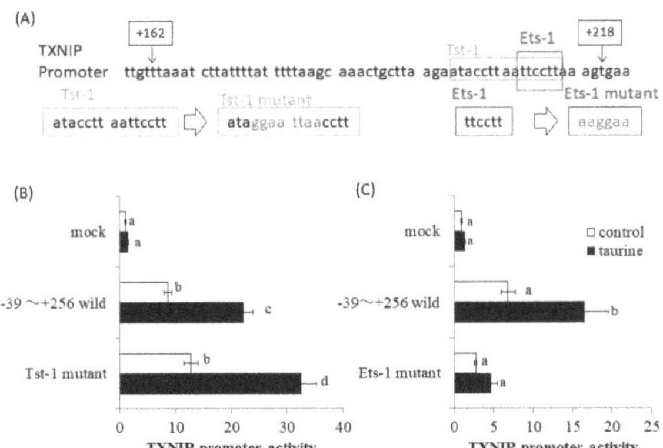

Figure 6. TXNIP promoter sequence (+162/+218) and the effect of taurine on TXNIP promoter activity by employing site-directed mutagenesis. (**A**) Human TXNIP promoter sequences between +162 and +218 are shown. The sequence of putative Tst-1 and Ets-1 response sequences and their mutants are shown, respectively. (**B**,**C**) Caco-2 cells were transfected with a reporter vector containing the partial promoter region of TXNIP (−39/+256; wild type, Tst-1 mutant, Ets-1 mutant) and then replaced with the medium containing 100 mM of taurine. After 24 h, the cells were subjected to the luciferase assay. Results are expressed as relative values with the control value of mock as 1. Each value is the mean ± S.E. ($n = 4$) and the abcd values indicated by different characters are significantly different from each other and the abcd values indicated by the same characters are not significantly different (Tukey's test; $p < 0.05$).

2.2. Effect of Taurine on Ets-1 Binding to TXNIP Promoter (ChIP Assay)

To confirm the enhancement of Ets-1 binding to the human TXNIP promoter region by taurine, a ChIP assay was performed. Caco-2 cells were cultured with 100 mM of taurine for 48 h and the cells were recovered. The lysate was immunoprecipitated using an anti-Ets-1 antibody. A real-time polymerase chain reaction (PCR) analysis was performed to determine whether the precipitated DNA contained the TXNIP promoter region with the Ets-1 response element. The results showed that taurine clearly increased Ets-1 binding to the TXNIP promoter region (Figure 7). We did not perform experimental verification of Tst-1 binding but only focused our attention on Ets-1.

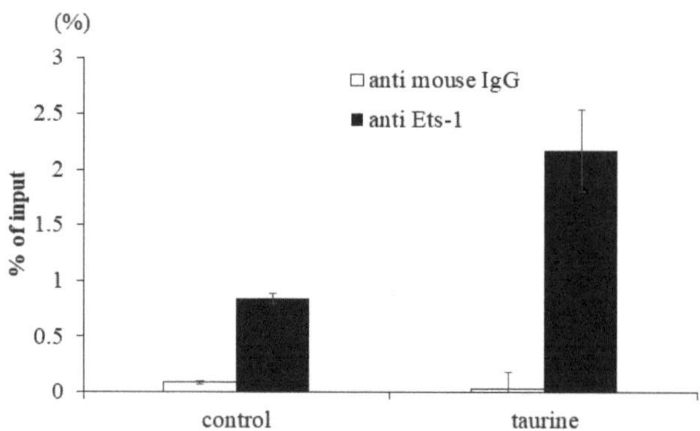

Figure 7. Effect of taurine on the binding of Ets-1 to the TXNIP promoter. Caco-2 cells were cultured in medium containing 100 mM of taurine. After 48 h, a ChIP assay was performed as described in Materials and Methods. The results are expressed as relative values with 100% being the DNA value before immunoprecipitation. Each value represents the mean ± S.E. (n = 2).

2.3. Effect of Taurine on Ets-1 Activation (Phosphorylation)

We also examined whether taurine activates Ets-1 or not. Ets-1 is activated by Thr38 [23]. Caco-2 cells were incubated with 100 mM of taurine for 3 h and the cell lysate was recovered and used for western blot analysis. Figure 8 shows that taurine increased the protein expression of phosphorylated Ets-1, suggesting that taurine activates Ets-1.

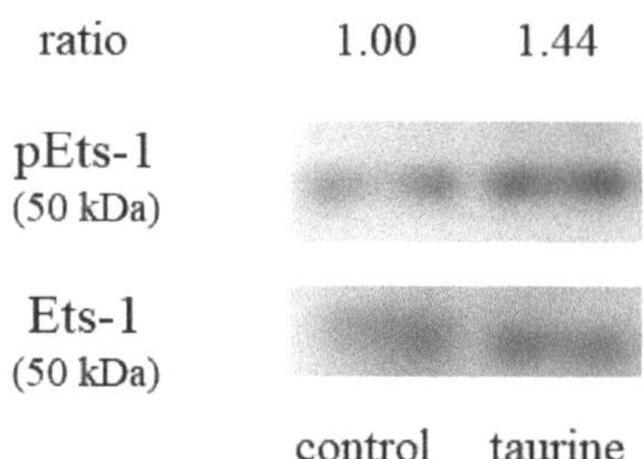

Figure 8. Effect of taurine on the phosphorylation (activation) of Ets-1. Caco-2 cells were cultured in a medium containing 100 mM of taurine for 3 h. The cell lysate was recovered and used for the western blot analysis. The bands were quantified using an image gauge.

2.4. The Involvement of MAP Kinase Family on Taurine-Induced Induction of TXNIP mRNA

To further elucidate the signaling pathway involved in taurine-induced TXNIP induction, we examined the effect of MAPK inhibitors on taurine TXNIP mRNA expression. The results showed that PD98059, an ERK1/2 pathway inhibitor, significantly suppressed the taurine-induced increase in TXNIP mRNA, whereas the p38 (SB203580) and JNK (SP600125) inhibitors did not (Figure 9).

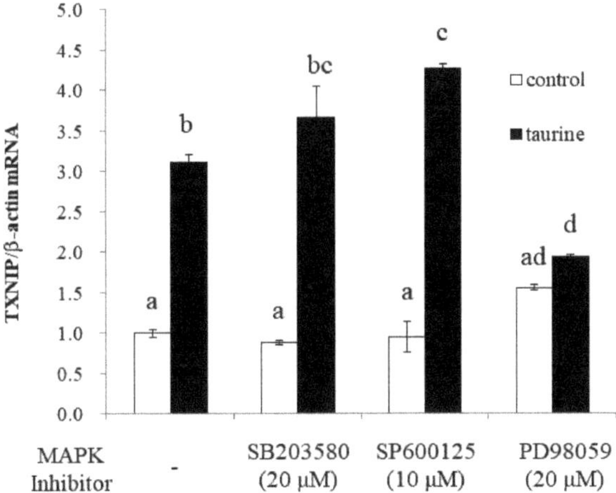

Figure 9. Effect of MAP kinase inhibitors on the taurine-induced increase in TXNIP mRNA in Caco-2 cells. Caco-2 cells were pretreated with each of the three MAPK inhibitors for 2 h and then cultured in medium containing 100 mM of taurine and each inhibitor. After 48 h, RNA was extracted and used for real-time PCR. The results are expressed as relative values, with the control value without the inhibitor as 1. Each value is the mean ± S.E. (n = 3) and the [abcd] values indicated by different characters are significantly different from each other, and the [abcd] values indicated by the same characters are not significantly different (Tukey's test; $p < 0.05$).

2.5. *Effect of Taurine on ERK1/2 Activation in Caco-2 Cells*

Next, to assess whether ERK1/2 was activated by taurine, the western blot analysis was performed. ERK1/2 was phosphorylated after 3 or 24 h of incubation with 100 mM of taurine (Figure 10). These results indicated that taurine activated ERK1/2 in Caco-2 cells.

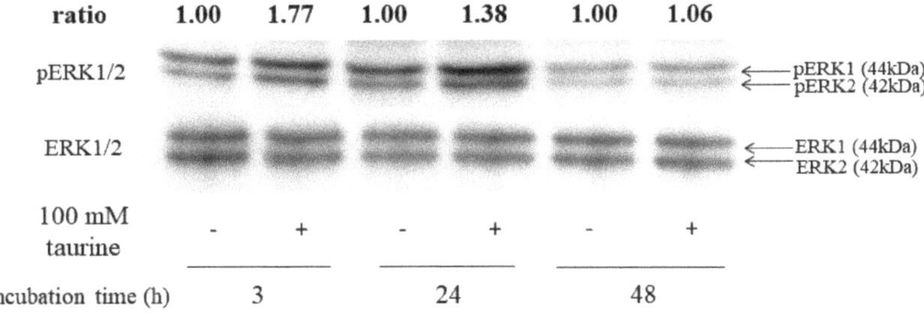

Figure 10. Effect of taurine on the phosphorylation (activation) of ERK1/2. Caco-2 cells were cultured in medium containing 100 mM of taurine for 3, 24, and 48 h. The cell lysate was recovered and used for the western blot analysis. The bands were quantified using an image gauge. Changes in expression are presented as relative values, with the expression level in the control at each treatment time as 1.

2.6. *Involvement of ERK Signaling Pathway in Taurine-Induced Enhancement of Transcriptional Activity of TXNIP*

Finally, we examined the effect of PD98059 on the taurine-induced increase in TXNIP promoter activity (+200/+218). PD98059 significantly suppressed the taurine-induced increase in TXNIP promoter activity (Figure 11). This result suggests that taurine activates the TXNIP promoter activity via the ERK cascade.

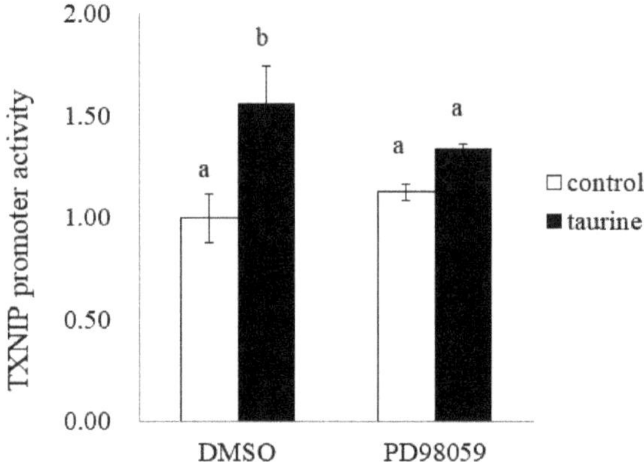

Figure 11. Effect of ERK1/2 cascade inhibitor on the taurine-induced increase in TXNIP promoter. After transfection with a reporter vector containing the promoter region of TXNIP (+200/+218), Caco-2 cells were pretreated with 50 μM PD98059 for 2 h and then cultured in medium containing 100 mM of taurine and the inhibitor. After 24 h, the cells were subjected to a luciferase assay. The results are expressed as relative values, with the control value without inhibitor as 1. Each value is the mean ± S.E. (n = 3) and the ^{ab} values indicated by different characters are significantly different from each other and the ^{ab} values indicated by the same characters are not significantly different (Tukey's test; $p < 0.05$).

3. Discussion

In the present study, we performed a detailed analysis of the mechanisms underlying the taurine-induced enhancement of TXNIP transcriptional activity. Our results demonstrate that taurine activates (phosphorylates) the transcription factor Ets-1 via the ERK signaling pathway and that the activated Ets-1 enhances the transcriptional activity of TXNIP by binding to the response sequence at +200/+218 in the TXNIP promoter region.

A total of 100 mM of taurine increases the TXNIP promoter activity (+200/+218) about two-fold in Figure 4, whereas taurine increases the promoter activity (+200/+218) about 1.3 fold in Figure 5. The difference in the induction ratio of luciferase activity by taurine between Figures 4 and 5 is thought to be due to the difference in the condition of Caco-2 cells. Caco-2 cells are used as an intestinal epithelial model, but this cell line is heterogeneous and it is well known that the cell characteristics and cell response are often changed during passage [24,25]. Although the induction ratio is different, the significant increase of TXNIP promoter activity (+200/+218) by taurine is observed in both Figures 4 and 5.

The deletion analysis of the TXNIP promoter suggests that the response sequences of the transcription factors Tst-1 and Ets-1 in the 200/+218 region of the TXNIP promoter were the candidates for taurine response sequences. However, there have been no reports of promoter analyses on TXNIP in region 3′ from −39, and it is unknown what kind of transcription factor response sequences exist in this region. Therefore, we analyzed the sequences using JASPAR, a database of predicted transcription factor-binding sequences. These results revealed the presence of a putative Tst-1 transcription response sequence and an Ets-1 transcription factor response sequence in +210/+215 of the TXNIP promoter.

Tst-1 belongs to a gene family with a Pit-Oct-Unc (POU) domain, and is reported to be expressed in the brain [26], and represses the expression of the P0 gene, a cell surface adhesion molecule [26], and specifically induces gene expression of the nicotinic acetylcholine receptor (nAchR) subunit 3 [27]. However, Tst-1 has only been examined in the brain and not much information is available about its role in other tissues. Ets-1 is a member of the E26 transformation-specific (Ets) family, which contains a helix-turn-helix DNA-binding region and has been reported to bind to the GGAA/T motif of DNA [28]. The Ets family is reported to be a family of transcription factors involved in diverse physiological functions, such as cell proliferation, differentiation, and apoptosis [29], and Ets-1 is known to be expressed in many tissues [30]. Therefore, we constructed luciferase vectors with mutations in the response sequences of each transcription factor (Tst-1 and Ets-1) and proceeded with the analysis.

Mutation analysis suggests the involvement of Ets-1 in taurine-induced enhancement of TXNIP transcription. Subsequently, we examined whether taurine activates Ets-1 and found that taurine addition activated (phosphorylated) Ets-1 (Figure 8). Considering that Ets-1 is reported to be activated by ERK [31], we hypothesized that taurine phosphorylates the transcription factor Ets-1 via ERK and enhances TXNIP transcriptional activity. A ChIP assay was performed on Caco-2 cells cultured in taurine-containing medium as shown in Figure 7; the results showed that taurine enhanced Ets-1 binding to the TXNIP promoter region. Notably, the binding of Ets-1 to the TXNIP promoter region was observed even in controls, suggesting that Ets-1 is involved in TXNIP transcriptional activity, even in the steady-state. This result is consistent with the results of the mutation analysis shown in Figure 6. A previous study reported that Ets-1 binds to the TXNIP promoter region at a steady-state in the regulation of TXNIP expression in pancreatic cells [32]. Further, Ets-1 has been shown to bind to the Flvcr1 promoter region in Caco-2 cells [33].

These findings suggest that taurine activates transcription factor Ets-1 via ERK and binds to the TXNIP promoter region, resulting in increased TXNIP transcriptional activity. The involvement of Ets-1 in the transcriptional activity of TXNIP has previously been reported in studies that observed induction of TXNIP by synthetic retinoids in osteosarcoma cells [34] and in the regulation of TXNIP expression in pancreatic cells [32]. However, considering that these two previous studies have shown that Ets-1 binds to the response sequence in the TXNIP promoter $-300/-400$ region, and our results that taurine-induced enhancement of TXNIP transcriptional activity occurs in the TXNIP promoter +200/+218 region, it appears that taurine may activate Ets-1 through multiple mechanisms. In addition to TXNIP, biomolecules that Ets-1 induces include connective tissue growth factor (CTGF/CCN2), a factor that promotes bone formation [35,36], vascular endothelial growth factor (VEGF) [37], parathyroid hormone-related peptide (PTHrP) [38], and p16 [39]. In addition, Ets-1 interacts with other transcription factors such as prox1, SP-1, and AP-1 to induce their expression [37,38]. Therefore, we cannot rule out the involvement of factors other than Ets-1 in the enhancement of TXNIP transcriptional activity by taurine.

Ets-1 is expressed in all tissues during embryonic development in mice and is reported to be crucial for morphogenesis [40]. In this process, Ets-1 plays a particularly important role in angiogenesis, and it has been reported that double-mutant mice with Ets-2 die in the early stages of angiogenesis [41]. As Ets-1 is involved in the formation of many tissues, it is likely that Ets-1 aids the key function of taurine during fetal, neonatal, and infant life. It has also been reported that natural killer cells do not mature in Ets-1-deficient mice [42]. Thus, Ets-1 appears to be involved in various physiological functions and likely regulates the diverse physiological effects of taurine. Our study elucidated the mechanism of taurine-induced enhancement of TXNIP expression and indicates that this occurs via the ERK pathway. However, the mechanism by which taurine activates ERK is not clear. Upstream signals of the ERK pathway include Ras/Raf/MEK/ERK [43], Ras/PKC/MEK/ERK [44], PKC/Ras/MEK/ERK [45], and PKC/MEK/ERK without Ras and Raf [46]. Since the PKC pathway is involved in three of these four pathways, future studies should examine the effect of the PKC inhibitor Ro318220 on taurine-induced increases in TXNIP mRNA expression. As PKC has also been reported to inhibit TAUT activity [47], it is possible that the cellular uptake of taurine is affected by it, though this possibility needs to be tested. Ras is involved in the activation of Ets-1 [48] and Ets-1 is activated by the Ras/Raf/MEK pathway, particularly in the induction of p16 expression [39]. Therefore, it is likely that taurine activates Ras, leading to the activation of ERK. Consequently, to clarify the upstream factors of the ERK pathway, it is necessary to first examine in detail whether taurine activates PKC and Ras. Further, it is reported that the inhibition of the ERK1/2 pathway by PD98059 in vitro leads to compensatory upregulation of the PI3K/Akt signaling pathway [49]. Therefore, it cannot be ruled out that PI3K/Akt activation is involved in this TXNIP induction. However, we recently found that taurine increases TXNIP mRNA expression in human hepatic HepG2 cells as well as in Caco-2 cells and further revealed that LY294002, a specific PI3K inhibitor, had no significant effect on taurine-induced increase in TXNIP mRNA (Figure A1). Therefore, it is not likely that PI3K/Akt signaling pathway is involved in this regulation.

Caco-2 cells are thought to recognize intracellular or extracellular taurine via unknown taurine receptors or binding proteins, leading to ERK-Ets-1 activation. In the brain, taurine binds to $GABA_A$ [50–52] and glycine [52,53] receptors and functions as an agonist. However, its binding affinity to the receptors is weaker than that of GABA and glycine. There have been no reports of binding to these receptors in tissues other than cranial nerves. Furthermore, taurine, but neither GABA nor glycine, contributes to the regulation of TAUT expression in intestinal epithelial cells [5]. TAUT receptors are expressed in the plasma membrane [54] and are activated by β-alanine [55]. Noticeably, in the present study, mRNA expression of TXNIP was not induced by β-alanine [9]; therefore, TAUT receptors are unlikely to be involved in the taurine-induced upregulation of TXNIP transcription. Furthermore, the expression of TAUT receptors has been reported only in neural tissues, and the DNA microarray results from our laboratory showed that their expression was absent from our cell cultures [9], suggesting that they are not expressed in Caco-2 cells. These findings suggest that there may be other receptors in cells that specifically recognize taurine; however, these receptors are yet to be identified.

In the present study, we revealed that taurine induced the transcriptional activity of TXNIP via the ERK and Ets-1 signaling pathways. These findings will hopefully contribute to the elucidation of the mechanisms underlying the diverse physiological effects of taurine and the discovery of new functions of taurine.

4. Materials and Methods

4.1. Materials

The Caco-2 cell line, derived from human colon cancer tissue, was purchased from the American Type Culture Collection (ATCC, Manassas, VA, USA). Dulbecco's Modified Eagle's Medium (DMEM) was purchased from Wako Pure Chemicals (Osaka, Japan). Fetal bovine serum (FBS) was purchased from Sigma-Aldrich (St. Louis, MO, USA). Penicillin-streptomycin (10,000 U/mL and 10 mg/mL in 0.9% sodium chloride) and non-essential amino acids (NEAA) were purchased from Gibco (Gaithersburg, MD, USA). The lipofectamine reagent was purchased from Invitrogen (Carlsbad, CA, USA), and the ExScript RT reagent kit and SYBR Premix Ex Taq for real-time PCR were obtained from Takara Bio (Otsu, Japan). All other chemicals used were of reagent grade and commercially available.

4.2. Cell Culture

Caco-2 cells were cultured in a medium consisting of DMEM, 10% FBS (*v*/*v*), 1% NEAA (*v*/*v*), 100 U/mL of penicillin, and 100 μg/mL of streptomycin. The cells were incubated at 37 °C in a humidified atmosphere containing 5% CO_2. The culture medium was replaced every other day. After reaching confluence, the cells were trypsinized with 0.1% trypsin and 0.02% EDTA in PBS and then subcultured. Caco-2 cells used in this study were between passages 35 and 79.

4.3. Plasmid Construct

The human TXNIP reporter vector containing the human TXNIP promoter region (−1299/+256) was inserted into the pGL4-basic vector (Promega, Madison, WI, USA) as previously described [10]. The human TXNIP promoter region (−109/+256, −39/+256, −39/+142, −39/+65, −1299/+142, and +122/+256, respectively) was cloned from the pGL4 reporter vector containing the TXNIP promoter (−1299/+256) by PCR. The PCR product was inserted into the pGL4-basic vector or pGL4-Promoter vector by digesting *Kpn*I and *Hind*III, and the restriction enzyme sequence was introduced into the primers.

Primers used for each reporter vector are listed in Table A1. The human TXNIP reporter vector containing the human TXNIP promoter region (+122/+178, +162/+218, +211/+256, +174/+191, +187/+204, and +200/+218) was constructed by ligating the pGL4-Promoter vector with each pair of oligonucleotides and annealing as listed in Tables A2 and A3.

4.4. *Transfection and Reporter Assay*

Caco-2 cells grown in a 24-well plate to 80% confluency were transiently transfected with 1 μg of the reporter vector and 0.05 μg of pRL-CMV using a lipofectamine reagent. Cells treated with or without 100 mM of taurine for 24 h were washed with PBS and lysed with Passive Lysis Buffer (Promega, Madison, WI, USA). Luciferase activity was measured using the dual-luciferase reporter assay (Promega, Madison, WI, USA) and an LB9507 Lumet luminometer (Berthold Technologies, Bad Wildbad, Germany).

4.5. *Real-Time PCR Analysis*

Total RNA was extracted from Caco-2 cells cultured with or without taurine using Isogen (Nippon Gene, Tokyo, Japan), according to the manufacturer's instructions. Reverse transcription of the RNA was performed using the PrimeScript RT Reagent Kit (TAKARA, Shiga, Japan), and first-strand cDNA was prepared from 0.5 μg of total RNA) and amplified using a SYBR Green Kit (TAKARA). The real-time PCR denaturation temperature was at 95 °C for 15 min, followed by 40 cycles of denaturation at 95 °C, at 60 °C for 15 s, and an extension at 72 °C for 10 s. The primer sequences were as follows: human TXNIP, 5′-ACGCTTCTTCTGGAAGACCA-3′ (forward), and 5′-AAGCTCAAAGCCGAACTTGT-3′ (reverse); β-actin, 5′-CCAGCACAATGAAGATCAAGA-3′ (forward) and 5′- AGAAAGGGT GTAACGCAACTAA-3′ (reverse). Real-time PCR was run on a LightCycler (Roche Applied Sciences, Penzberg, Germany).

4.6. *Chip Assay*

Caco-2 cells were cultured in 6-well plates and incubated with taurine for 48 h. The cells were then homogenized, and the nuclear extract was prepared using the ChIP assay. The ChIP assay was performed using ChIP-IT Express (Active Motif, Carlsbad, CA, USA), according to the manufacturer's instructions. To quantify the number of DNA fragments containing the TXNIP promoter region bound by Ets-1 protein, real-time PCR was performed. The primer sequences were as follows: human TXNIP promoter, 5′-TCGGATCTTTCTCCAGCAAT-3′ (forward), and 5′-AAATCGAGGAAACCCCTTTG-3′ (reverse).

4.7. *Western Blot Analysis*

Caco-2 cells were cultured in 6-well plates with or without taurine for several hours. The cells were collected with a cell scraper and suspended in lysis buffer (50 mM HEPES (pH 7.5), 150 mM NaCl, 1% NP-40, 0.5% sodium deoxycholate, 1 mM $NaVO_3$, 50 mM NaF, 20 mM β-glycerophosphate, 0.1% inhibitor cocktail (Sigma, St. Louis, MO, USA). The cell homogenate was centrifuged at 20,000× *g* for 10 min at 4 °C. The supernatant was used for western blot analysis as described previously [11]. The protein assay was performed using the Bio-Rad Protein Assay Solution (Bio-Rad, Hercules, CA, USA). The primary antibodies used were rabbit anti-human p-Ets-1 (Thr38) (Abcam, Cambridge, UK) and anti-human Ets-1 (Santa Cruz, Dallas, TX, USA). Rabbit anti-human p-ERK1/2 (Thr202/Tyr204) antibody (Santa Cruz, Dallas, TX, USA) and rabbit anti-human ERK1/2 antibody (Santa Cruz, Dallas, TX, USA) were used as the primary antibodies. The secondary antibody was goat anti-rabbit IgG antibody conjugated to horseradish peroxidase (Amersham PLC, Bucks, UK). The bound antibodies were analyzed using an ECL plus western blotting detection system (GE Healthcare, Boston, MA, USA) and a Lumino Image Analyzer (ImageQuant LAS-4000 miniPR; Cytiva, Krefeld, Germany). Original, whole membrane images were provided for the review process and are available from the journal upon request.

4.8. Statistical Analysis

Data are expressed as mean ± SE. Statistical comparisons were performed using the Student's *t*-test, Dunnett's test, or Tukey's test.

Author Contributions: H.S. (Hideo Satsu), Y.G. and H.S. (Hana Shimanaka) conceived and designed the experiments; H.S. (Hideo Satsu), Y.G., H.S. (Hana Shimanaka) and K.W. performed the experiments and analyzed the data; S.W., S.-J.P. and K.N. conceived and performed the informatic analysis; H.S. (Hideo Satsu), Y.G., H.S. (Hana Shimanaka), M.I., S.M. and M.S. discussed the data; and H.S. (Hideo Satsu) wrote the paper. All authors have read and agreed to the published version of the manuscript.

Funding: This work was funded in part by a research grant from Taisho Pharmaceutical Co., Ltd.

Institutional Review Board Statement: Not applicable.

Informed Consent Statement: Not applicable.

Data Availability Statement: The data supporting the findings of this study are available from the corresponding author, H.S., upon reasonable request due to restrictions on privacy.

Acknowledgments: We acknowledge Mika Takahashi for the technical support.

Conflicts of Interest: The authors declare no conflict of interest.

Appendix A

Table A1. Primers used for constructing each TXNIP promoter vector.

Region		Primer Sequences (5′→3′)
−1299/+256	forward	CCGGTACCCCAACAAAGAATGAAGAGAGAG
	reverse	GCAAGCTTCTCCAAATCGAGGAAACCC
−109/+256	forward	CCGGTACCAGCCAATGGGAGGGATG
	reverse	GCAAGCTTCTCCAAATCGAGGAAACCC
−39/+256	forward	CCGGTACCCGGGCTACTATATAGAGACG
	reverse	GCAAGCTTCTCCAAATCGAGGAAACCC
−39/+142	forward	CCGGTACCCGGGCTACTATATAGAGACG
	reverse	GCAAGCTTCTAGGTTTTCGAAAAGGCGCC
−39/+65	forward	CCGGTACCCGGGCTACTATATAGAGACG
	reverse	GCAAGCTTCCCCAATTGCTGGAGAAAAG
−1299/+142	forward	CCGGTACCCCAACAAAGAATGAAGAGAGAG
	reverse	GCAAGCTTCTAGGTTTTCGAAAAGGCGCC
+122/+256	forward	CCGGTACCGGCGCCTTTTCGAAAACCTAG
	reverse	GCAAGCTTCTCCAAATCGAGGAAACCC

Table A2. Partial TXNIP promoter sequence (+122/+256) of oligonucleotide for annealing.

Region		Primer Sequences (5′→3′)
+122/+178	forward	cggcgccttttcgaaaacctagtagttaatattcatttgtttaaatcttattttata
	reverse	agcttataaaataagatttaaacaaatgaatattaactactaggttttcgaaaaggcgccggtac
+162/+218	forward	cttaaatcttattttatttttaagctcaaactgcttaagaataccttaattccttaaaga
	reverse	agcttctttaaggaattaaggtattcttaagcagtttgagcttaaaaataaaataagatttaaggtac
+211/+256	forward	cccttaaagtgaaataatttttgcaaaggggtttcctcgatttggaga
	reverse	agcttctccaaatcgaggaaacccctttgcaaaaattatttcactttaaggggtac

Table A3. Partial TXNIP promoter (+174/+218) sequence of oligonucleotide for annealing.

Region		
+174/+191	forward	cttatttttaagctcaaaca
	reverse	agcttgtttgagcttaaaaataaggtac
+187/+204	forward	ccaaactgcttaagaataca
	reverse	agcttgtattcttaagcagtttgggtac
+200/+218	forward	caataccttaattccttaaaa
	reverse	agctttttaaggaattaaggtattggtac

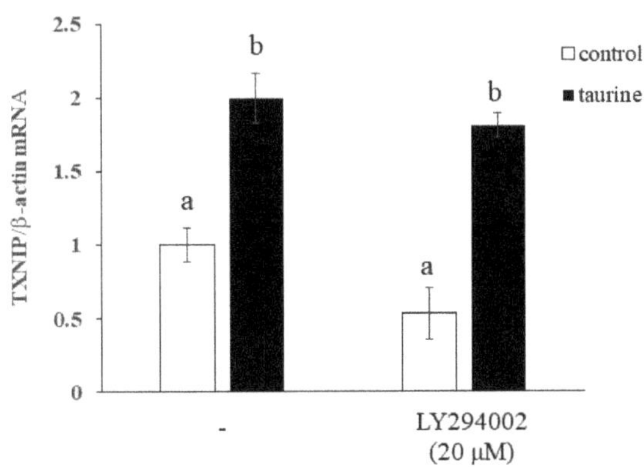

Figure A1. Effect of the PI3K inhibitor on taurine-induced increase in TXNIP mRNA in human hepatic HepG2 cells. HepG2 cells were pretreated with 20 μM LY294002 for 2 h and then cultured in medium containing 100 mM of taurine and the inhibitor. After 48 h, RNA was extracted and used for real-time PCR. The results are expressed as relative values, with the control value without LY294002 as 1. Each value is the mean ± S.E. (n = 4), and the ^{ab} values indicated by different characters are significantly different from each other and the ^{ab} values indicated by the same characters are not significantly different (Tukey's test; $p < 0.05$).

References

1. Huxtable, R.J. Physiological actions of taurine. *Physiol. Rev.* **1992**, *72*, 101–163. [CrossRef] [PubMed]
2. Schaffer, S.W.; Azuma, J.; Takahashi, K.; Mozaffari, M. Why is taurine cytoprotective? *Adv. Exp. Med. Biol.* **2003**, *526*, 307–321. [PubMed]
3. Sturman, J.A. Taurine in development. *J. Nutr.* **1988**, *118*, 1169–1176. [CrossRef] [PubMed]
4. Tochitani, S. Functions of Maternally-Derived Taurine in Fetal and Neonatal Brain Development. *Adv. Exp. Med. Biol.* **2019**, *975*, 17–25.
5. Satsu, H.; Watanabe, H.; Arai, S.; Shimizu, M. Characterization and regulation of taurine transport in Caco-2, human intestinal cells. *J. Biochem.* **1997**, *121*, 1082–1087. [CrossRef] [PubMed]
6. Satsu, H.; Miyamoto, Y.; Shimizu, M. Hypertonicity stimulates taurine uptake and transporter gene expression in Caco-2 cells. *Biochim. Biophys. Acta* **1999**, *1419*, 89–96. [CrossRef]
7. Mochizuki, T.; Satsu, H.; Shimizu, M. Tumor necrosis factor alpha stimulates taurine uptake and transporter gene expression in human intestinal Caco-2 cells. *FEBS Lett.* **2002**, *517*, 92–96. [CrossRef]
8. Zhao, Z.; Satsu, H.; Fujisawa, M.; Hori, M.; Ishimoto, Y.; Totsuka, M.; Nambu, A.; Kakuta, S.; Ozaki, H.; Shimizu, M. Attenuation by dietary taurine of dextran sulfate sodium-induced colitis in mice and of THP-1-induced damage to intestinal Caco-2 cell monolayers. *Amino Acids* **2008**, *35*, 217–224. [CrossRef]
9. Gondo, Y.; Satsu, H.; Ishimoto, Y.; Iwamoto, T.; Shimizu, M. Effect of taurine on mRNA expression of thioredoxin interacting protein in Caco-2 cells. *Biochem. Biophys. Res. Commun.* **2012**, *426*, 433–437. [CrossRef]
10. Kaimul, A.M.; Nakamura, H.; Masutani, H.; Yodoi, J. Thioredoxin and thioredoxin-binding protein-2 in cancer and metabolic syndrome. *Free Radic. Biol. Med.* **2007**, *43*, 861–868. [CrossRef]
11. Kim, S.Y.; Suh, H.W.; Chung, J.W.; Yoon, S.R.; Choi, I. Diverse functions of VDUP1 in cell proliferation, differentiation, and diseases. *Cell Mol. Immunol.* **2007**, *4*, 345–351. [PubMed]

12. Andres, A.M.; Ratliff, E.P.; Sachithanantham, S.; Hui, S.T. Diminished AMPK signaling response to fasting in thioredoxin-interacting protein knockout mice. *FEBS Lett.* **2011**, *585*, 1223–1230. [CrossRef]
13. Donnelly, K.L.; Margosian, M.R.; Sheth, S.S.; Lusis, A.J.; Parks, E.J. Increased lipogenesis and fatty acid reesterification contribute to hepatic triacylglycerol stores in hyperlipidemic Txnip-/- mice. *J. Nutr.* **2004**, *134*, 1475–1480. [CrossRef] [PubMed]
14. Hui, S.T.; Andres, A.M.; Miller, A.K.; Spann, N.J.; Potter, D.W.; Post, N.M.; Chen, A.Z.; Sachithanantham, S.; Jung, D.Y.; Kim, J.K.; et al. Txnip balances metabolic and growth signaling via PTEN disulfide reduction. *Proc. Natl. Acad. Sci. USA* **2008**, *105*, 3921–3926. [CrossRef] [PubMed]
15. Oka, S.; Liu, W.; Masutani, H.; Hirata, H.; Shinkai, Y.; Yamada, S.; Yoshida, T.; Nakamura, H.; Yodoi, J. Impaired fatty acid utilization in thioredoxin binding protein-2 (TBP-2)-deficient mice: A unique animal model of Reye syndrome. *FASEB J.* **2006**, *20*, 121–123. [CrossRef]
16. Sheth, S.S.; Castellani, L.W.; Chari, S.; Wagg, C.; Thipphavong, C.K.; Bodnar, J.S.; Tontonoz, P.; Attie, A.D.; Lopaschuk, G.D.; Lusis, A.J. Thioredoxin-interacting protein deficiency disrupts the fasting-feeding metabolic transition. *J. Lipid Res.* **2005**, *46*, 123–134. [CrossRef]
17. van Greevenbroek, M.M.; Vermeulen, V.M.; Feskens, E.J.; Evelo, C.T.; Kruijshoop, M.; Hoebee, B.; van der Kallen, C.J.; de Bruin, T.W. Genetic variation in thioredoxin interacting protein (TXNIP) is associated with hypertriglyceridaemia and blood pressure in diabetes mellitus. *Diabet. Med.* **2007**, *24*, 498–504. [CrossRef]
18. Chutkow, W.A.; Patwari, P.; Yoshioka, J.; Lee, R.T. Thioredoxin-interacting protein (Txnip) is a critical regulator of hepatic glucose production. *J. Biol. Chem.* **2008**, *283*, 2397–2406. [CrossRef]
19. Ferreira, N.E.; Omae, S.; Pereira, A.; Rodrigues, M.V.; Miyakawa, A.A.; Campos, L.C.; Santos, P.C.; Dallan, L.A.; Martinez, T.L.; Santos, R.D.; et al. Thioredoxin interacting protein genetic variation is associated with diabetes and hypertension in the Brazilian general population. *Atherosclerosis* **2012**, *221*, 131–136. [CrossRef]
20. Yoshioka, J.; Schulze, P.C.; Cupesi, M.; Sylvan, J.D.; MacGillivray, C.; Gannon, J.; Huang, H.; Lee, R.T. Thioredoxin-interacting protein controls cardiac hypertrophy through regulation of thioredoxin activity. *Circulation* **2004**, *109*, 2581–2586. [CrossRef]
21. Takahashi, Y.; Masuda, H.; Ishii, Y.; Nishida, Y.; Kobayashi, M.; Asai, S. Decreased expression of thioredoxin interacting protein mRNA in inflamed colonic mucosa in patients with ulcerative colitis. *Oncol. Rep.* **2007**, *18*, 531–535. [CrossRef] [PubMed]
22. Satsu, H.; Gondo, Y.; Shimanaka, H.; Watari, K.; Fukumura, M.; Shimizu, M. Effect of taurine on cell function via TXNIP induction in Caco-2 cells. *Adv. Exp. Med. Biol.* **2019**, *1155*, 163–172. [PubMed]
23. Garrett-Sinha, L.A. Review of Ets1 structure, function, and roles in immunity. *Cell. Mol. Life Sci.* **2013**, *70*, 3375–3390. [CrossRef]
24. Yu, H.; Cook, T.J.; Sinko, P.J. Evidence for diminished functional expression of intestinal transporters in Caco-2 cell monolayers at high passages. *Pharm. Res.* **1997**, *14*, 757–762. [CrossRef] [PubMed]
25. Steffansen, B.; Pedersen, M.D.L.; Laghmoch, A.M.; Nielsen, C.U. SGLT1-Mediated Transport in Caco-2 Cells Is Highly Dependent on Cell Bank Origin. *J. Pharm. Sci.* **2017**, *106*, 2664–2670. [CrossRef]
26. He, X.; Gerrero, R.; Simmons, D.M.; Park, R.E.; Lin, C.J.; Swanson, L.W.; Rosenfeld, M.G. Tst-1, a member of the POU domain gene family, binds the promoter of the gene encoding the cell surface adhesion molecule P0. *Mol. Cell. Biol.* **1991**, *11*, 1739–1744.
27. Yang, X.; McDonough, J.; Fyodorov, D.; Morris, M.; Wang, F.; Deneris, E.S. Characterization of an acetylcholine receptor alpha 3 gene promoter and its activation by the POU domain factor SCIP/Tst-1. *J. Biol. Chem.* **1994**, *269*, 10252–10264. [CrossRef]
28. Oikawa, T.; Yamada, T. Molecular biology of the Ets family of transcription factors. *Gene* **2003**, *303*, 11–34. [CrossRef]
29. Sementchenko, V.I.; Watson, D.K. Ets target genes: Past, present and future. *Oncogene* **2000**, *19*, 6533–6548. [CrossRef]
30. Dittmer, J. The biology of the Ets1 proto-oncogene. *Mol. Cancer* **2003**, *2*, 29. [CrossRef]
31. Roskoski, R., Jr. ERK1/2 MAP kinases: Structure, function, and regulation. *Pharmacol. Res.* **2012**, *66*, 105–143. [CrossRef] [PubMed]
32. Luo, Y.; He, F.; Hu, L.; Hai, L.; Huang, M.; Xu, Z.; Zhang, J.; Zhou, Z.; Liu, F.; Dai, Y.S. Transcription factor Ets1 regulates expression of thioredoxin-interacting protein and inhibits insulin secretion in pancreatic beta-cells. *PLoS ONE* **2014**, *9*, e99049.
33. Fiorito, V.; Neri, F.; Pala, V.; Silengo, L.; Oliviero, S.; Altruda, F.; Tolosano, E. Hypoxia controls Flvcr1 gene expression in Caco2 cells through HIF2alpha and ETS1. *Biochim. Biophys. Acta* **2014**, *1839*, 259–264. [CrossRef]
34. Hashiguchi, K.; Tsuchiya, H.; Tomita, A.; Ueda, C.; Akechi, Y.; Sakabe, T.; Kurimasa, A.; Nozaki, M.; Yamada, T.; Tsuchida, S.; et al. Involvement of ETS1 in thioredoxin-binding protein 2 transcription induced by a synthetic retinoid CD437 in human osteosarcoma cells. *Biochem. Biophys. Res. Commun.* **2010**, *391*, 621–626. [CrossRef] [PubMed]
35. Van Beek, J.P.; Kennedy, L.; Rockel, J.S.; Bernier, S.M.; Leask, A. The induction of CCN2 by TGFbeta1 involves Ets-1. *Arthritis Res. Ther.* **2006**, *8*, R36. [CrossRef]
36. Geisinger, M.T.; Astaiza, R.; Butler, T.; Popoff, S.N.; Planey, S.L.; Arnott, J.A. Ets-1 is essential for connective tissue growth factor (CTGF/CCN2) induction by TGF-beta1 in osteoblasts. *PLoS ONE* **2012**, *7*, e35258. [CrossRef]
37. Yoshimatsu, Y.; Yamazaki, T.; Mihira, H.; Itoh, T.; Suehiro, J.; Yuki, K.; Harada, K.; Morikawa, M.; Iwata, C.; Minami, T.; et al. Ets family members induce lymphangiogenesis through physical and functional interaction with Prox1. *J. Cell Sci.* **2011**, *124*, 2753–2762. [CrossRef]
38. Itoh, T.; Ando, M.; Tsukamasa, Y.; Akao, Y. Expression of BMP-2 and Ets1 in BMP-2-stimulated mouse pre-osteoblast differentiation is regulated by microRNA-370. *FEBS Lett.* **2012**, *586*, 1693–1701. [CrossRef]
39. Ohtani, N.; Zebedee, Z.; Huot, T.J.; Stinson, J.A.; Sugimoto, M.; Ohashi, Y.; Sharrocks, A.D.; Peters, G.; Hara, E. Opposing effects of Ets and Id proteins on p16INK4a expression during cellular senescence. *Nature* **2001**, *409*, 1067–1070. [CrossRef]

40. Kola, I.; Brookes, S.; Green, A.R.; Garber, R.; Tymms, M.; Papas, T.S.; Seth, A. The Ets1 transcription factor is widely expressed during murine embryo development and is associated with mesodermal cells involved in morphogenetic processes such as organ formation. *Proc. Natl. Acad. Sci. USA* **1993**, *90*, 7588–7592. [CrossRef]
41. Wei, G.; Srinivasan, R.; Cantemir-Stone, C.Z.; Sharma, S.M.; Santhanam, R.; Weinstein, M.; Muthusamy, N.; Man, A.K.; Oshima, R.G.; Leone, G.; et al. Ets1 and Ets2 are required for endothelial cell survival during embryonic angiogenesis. *Blood* **2009**, *114*, 1123–1130. [CrossRef] [PubMed]
42. Barton, K.; Muthusamy, N.; Fischer, C.; Ting, C.N.; Walunas, T.L.; Lanier, L.L.; Leiden, J.M. The Ets-1 transcription factor is required for the development of natural killer cells in mice. *Immunity* **1998**, *9*, 555–563. [CrossRef]
43. Harrisingh, M.C.; Lloyd, A.C. Ras/Raf/ERK signalling and NF1. *Cell Cycle* **2004**, *3*, 1255–1258. [CrossRef] [PubMed]
44. Zhao, Y.; Liu, J.; Li, L.; Liu, L.; Wu, L. Role of Ras/PKCzeta/MEK/ERK1/2 signaling pathway in angiotensin II-induced vascular smooth muscle cell proliferation. *Regul. Pept.* **2005**, *128*, 43–50. [CrossRef] [PubMed]
45. Lee, H.Y.; Crawley, S.; Hokari, R.; Kwon, S.; Kim, Y.S. Bile acid regulates MUC2 transcription in colon cancer cells via positive EGFR/PKC/Ras/ERK/CREB, PI3K/Akt/IkappaB/NF-kappaB and p38/MSK1/CREB pathways and negative JNK/c-Jun/AP-1 pathway. *Int. J. Oncol.* **2010**, *36*, 941–953.
46. Wen-Sheng, W. Protein kinase C alpha trigger Ras and Raf-independent MEK/ERK activation for TPA-induced growth inhibition of human hepatoma cell HepG2. *Cancer Lett.* **2006**, *239*, 27–35. [CrossRef]
47. Loo, D.D.; Hirsch, J.R.; Sarkar, H.K.; Wright, E.M. Regulation of the mouse retinal taurine transporter (TAUT) by protein kinases in Xenopus oocytes. *FEBS Lett.* **1996**, *392*, 250–254. [CrossRef]
48. Yang, B.S.; Hauser, C.A.; Henkel, G.; Colman, M.S.; Van Beveren, C.; Stacey, K.J.; Hume, D.A.; Maki, R.A.; Ostrowski, M.C. Ras-mediated phosphorylation of a conserved threonine residue enhances the transactivation activities of c-Ets1 and c-Ets2. *Mol. Cell. Biol.* **1996**, *16*, 538–547. [CrossRef]
49. Hu, C.; Huang, L.; Gest, C.; Xi, X.; Janin, A.; Soria, C.; Li, H.; Lu, H. Opposite regulation by PI3K/Akt and MAPK/ERK pathways of tissue factor expression, cell-associated procoagulant activity and invasiveness in MDA-MB-231 cells. *J. Hematol. Oncol.* **2012**, *5*, 16. [CrossRef]
50. Taber, K.H.; Lin, C.T.; Liu, J.W.; Thalmann, R.H.; Wu, J.Y. Taurine in hippocampus: Localization and postsynaptic action. *Brain Res.* **1986**, *386*, 113–121. [CrossRef]
51. del Olmo, N.; Bustamante, J.; del Río, R.M.; Solís, J.M. Taurine activates GABA(A) but not GABA(B) receptors in rat hippocampal CA1 area. *Brain Res.* **2000**, *864*, 298–307. [CrossRef]
52. Hussy, N.; Deleuze, C.; Pantaloni, A.; Desarménien, M.G.; Moos, F. Agonist action of taurine on glycine receptors in rat supraoptic magnocellular neurones: Possible role in osmoregulation. *J. Physiol.* **1997**, *502*, 609–621. [CrossRef] [PubMed]
53. Lewis, C.A.; Ahmed, Z.; Faber, D.S. A characterization of glycinergic receptors present in cultured rat medullary neurons. *J. Neurophysiol.* **1991**, *66*, 1291–1303. [CrossRef] [PubMed]
54. Whatley, V.J.; Harris, R.A. The cytoskeleton and neurotransmitter receptors. *Int. Rev. Neurobiol.* **1996**, *39*, 113–143. [PubMed]
55. Horikoshi, T.; Asanuma, A.; Yanagisawa, K.; Anzai, K.; Goto, S. Taurine and beta-alanine act on both GABA and glycine receptors in Xenopus oocyte injected with mouse brain messenger RNA. *Brain Res.* **1988**, *464*, 97–105. [PubMed]

MDPI

Review

The Role of Taurine in Skeletal Muscle Functioning and Its Potential as a Supportive Treatment for Duchenne Muscular Dystrophy

Caroline Merckx [1,*] and Boel De Paepe [1,2]

1 Department of Neurology, Ghent University Hospital, 9000 Ghent, Belgium; boel.depaepe@ugent.be
2 Neuromuscular Reference Center, Ghent University Hospital, 9000 Ghent, Belgium
* Correspondence: caroline.merckx@ugent.be

Abstract: Taurine (2-aminoethanesulfonic acid) is required for ensuring proper muscle functioning. Knockout of the taurine transporter in mice results in low taurine concentrations in the muscle and associates with myofiber necrosis and diminished exercise capacity. Interestingly, regulation of taurine and its transporter is altered in the mdx mouse, a model for Duchenne Muscular Dystrophy (DMD). DMD is a genetic disorder characterized by progressive muscle degeneration and weakness due to the absence of dystrophin from the muscle membrane, causing destabilization and contraction-induced muscle cell damage. This review explores the physiological role of taurine in skeletal muscle and the consequences of a disturbed balance in DMD. Its potential as a supportive treatment for DMD is also discussed. In addition to genetic correction, that is currently under development as a curative treatment, taurine supplementation has the potential to reduce muscle inflammation and improve muscle strength in patients.

Keywords: taurine; osmolytes; muscle; Duchenne Muscular Dystrophy; mdx

Citation: Merckx, C.; De Paepe, B. The Role of Taurine in Skeletal Muscle Functioning and Its Potential as a Supportive Treatment for Duchenne Muscular Dystrophy. *Metabolites* **2022**, *12*, 193. https://doi.org/10.3390/metabo12020193

Academic Editors: Cholsoon Jang, Teruo Miyazaki, Takashi Ito, Alessia Baseggio Conrado and Shigeru Murakami

Received: 21 January 2022
Accepted: 16 February 2022
Published: 19 February 2022

Publisher's Note: MDPI stays neutral with regard to jurisdictional claims in published maps and institutional affiliations.

1. Introduction

Taurine or 2-aminoethane-sulfonic acid is primarily a free occurring sulfur-containing amino acid. Unlike most other amino acids, it is not a building block for proteins, yet classifies as a conditionally essential amino acid that is abundant in excitable tissues such as brain, retina, heart, and skeletal muscle, where intracellular concentrations range from 20 to 70 mmol/kg. Taurine is either taken up from diet, for example from fish and meat, or can be synthesized from other amino acids such as cysteine or methionine. Taurine has versatile functions: it plays an important role in osmoregulation, acts as a stabilizer of the cell membrane and of proteins, has anti-oxidant and anti-inflammatory functions, regulates mitochondrial tRNA activities, is involved in calcium homeostasis, etc. [1–3].

In this review, we focused on the role of taurine in muscle disease, especially in Duchenne Muscular Dystrophy (DMD), a progressive muscle wasting disorder affecting approximately 1 per 5000 male births [4]. Muscle weakness is conspicuous in the hip- and pelvic area first, and later spreads to distal regions. Patients become wheelchair-dependent in their early teens and eventually require cardiac and respiratory care since the muscles of the heart and respiratory system are affected in a life-threatening manner. While awaiting curative treatments to enter the clinic, glucocorticoids are the standard of care, and can prolong life-expectancy of DMD patients [5,6]. Although the precise mechanism by which glucocorticoids slow down disease progression in DMD is not completely understood, its anti-inflammatory action might play a crucial role. However, the use of glucocorticoids negatively influences bone health, which is already impaired in DMD patients [7]. A comprehensive overview of emerging genetic therapies in DMD is provided in the review of Sun et al. [8]

The genetic cause of DMD is mutations in the dystrophin gene, located on chromosome X, which hampers the production of functional dystrophin protein. The latter is a key component of the dystrophin-associated protein complex (DAPC) that provides stability to muscle fibers during contraction and relaxation by connecting the intracellular actin cytoskeleton to the basal lamina [7]. Besides membrane stabilization, DAPC also fulfils a role in signal transduction. Dystrophic muscles encounter chronic inflammation, oxidative stress, and ischemia. Eventually these detrimental processes lead to loss of muscle mass and muscle fibrosis [7].

This review explores the physiological role of taurine in skeletal muscle and focuses on the consequences of its disturbed balance in DMD. The therapeutic potential of taurine as a dietary supplement for DMD will be scrutinized.

2. Involvement of Taurine in Physiological Skeletal Muscle Functioning

2.1. Knowledge Gained from Knockout Models

A first clue towards an important role for taurine in the muscle was its relatively high abundance. Proper insights regarding the function of taurine in physiological muscle function were acquired upon the generation of taurine transporter (TauT) knock-out (KO) mouse models and ablation of muscle taurine content. Lack of taurine impaired the conductance velocity of the muscle without affecting nerve conductance speed [2]. In addition, exercise performance was seriously hampered in TauT KO mice as shown by a significantly lower running speed. In a different experimental set-up, the total running distance was only 20% of the distance travelled by age-matched wild type (WT) mice [2,9,10]. Besides running tests, the reduced exercise capacity of TauT KO mice also became apparent during a weight-loaded swimming test that showed an 80% decrease in swimming time compared to WT [11]. This study reported structural changes in morphology of TauT KO muscle; however, Warskulat et al. hypothesized that hampered exercise performance was likely attributed to muscle dysfunction, resulting from taurine deficiency. In evidence, serum creatine kinase levels are increased in TauT KO and serum lactate levels were raised after exercise [2]. Some of the pathological characteristics of TauT KO models such as necrotic myofibers and reduced exercise capacity resemble the features of the mdx mouse model [12,13]. Another approach to evaluate the effect of taurine depletion is the use of guanodinoethane sulfonate (GES). GES, a taurine transporter antagonist, reduces muscle taurine content by 60% [14]. Previously taurine was shown to enhance calcium uptake and release in myofibers concurrently with an increase in force production, whereas myofibers derived from GES-treated mice showed reduced force production at relevant stimulation frequencies. Interestingly, fatigue was reportedly attenuated upon GES-treatment [14].

Furthermore, the lifespan of TauT KO male mice is significantly lower than those of WT mice, 511 and 686 days, respectively. Reduced life expectancy together with increased expression of p16INK4a, an indicator of senescence, in TauT KO mice allowed to hypothesize that taurine might be involved in aging [15,16]. Furthermore, it was suggested that taurine delays muscle-specific senescence including sarcopenia in tumor necrosis factor (TNF)-α stimulated L6 myogenic rat cells. In evidence, differences in the regulation of inflammation, autophagy, and apoptosis have been reported upon taurine treatment [17]. In addition, TNF-α stimulation of L6 myogenic rat cells hampered muscle differentiation, which could be restored by taurine. Presumably this effect was mediated through the PI3/AKT signaling pathway since myocyte enhancer factor-2 (MEF-2), a transcription factor involved in myogenic differentiation, was markedly decreased after knockdown of AKT [17]. Furthermore, expression of TauT increased during muscle differentiation and was even further enhanced upon binding of MEF-2 to the promotor region of TauT [18].

In conclusion, depletion of muscle taurine levels either through TauT KO or via pharmacological inhibition of the taurine transporter by GES alters force output and exercise performance. Therefore, taurine seems essential for the preservation of physiological muscle function. The sections below provide an overview of the main cellular processes in which taurine plays a role in relation to the muscle.

2.2. Taurine and Its Role in Osmotic Homeostasis

Exposure to hyperosmolar conditions can induce several detrimental cellular effects including interference with transcriptional and translational activity, induction of oxidative stress, DNA damage, and can even elicit apoptosis under certain conditions [19,20]. Thus, safeguarding the osmotic equilibrium is essential for ensuring cellular health. When cells are exposed to an environment high on NaCl, fluid is retracted from the intracellular compartment causing cellular shrinkage, molecular crowding, and increased ionic forces. In order to counteract these deleterious processes, the cell's response is a regulatory volume increase (RVI) that includes activation of inorganic ion transporters (e.g., $Na^+/K^+/2Cl^-$ cotransporter, the $Na^+/H^{+/-}$, and the Cl^-/HCO_3^- exchanger), allowing an influx of ions accompanied by osmotic uptake of water [19–23]. However, this condition is unfavorable over time due to increased intracellular ionic forces that could interact with macromolecules. Thus secondarily, accumulation of organic osmolytes (e.g., taurine) that replace inorganic electrolytes results in normalization of ionic strength whilst preserving cellular volume and protein stability [19,20]. Rather than stimulation of de novo synthesis, osmotic stress is most likely to enhance cellular import of taurine [20]. Both the designated TauT as well as the proton-coupled amino-acid transporter (PAT) 1 are capable of accumulating taurine in the cell [24]. However, TauT is considered as the principal transporter of taurine in muscle cells as evidenced by a 98% reduction of taurine content in a TauT KO mouse [10]. Transcription of TauT mRNA is upregulated under hypertonic conditions due to binding of Nuclear Factor of Activated T-cells 5 (NFAT-5) to the 5′ flank region of the TauT gene. NFAT-5, also known as tonicity responsive element binding protein (TonEBP) acts as a transcription factor of *SLC6A6*, the gene encoding TauT, and thus allows cellular accumulation of taurine [25,26].

Exercise can affect the osmotic balance in muscle fibers. Muscle subjected to an intensive exercise protocol resulted in increased myofiber volume, cross-sectional area, and water concentration by more than 15%, indicative for muscle fiber swelling [27,28]. The rise in myofiber water content can be partially explained by water production in cellular metabolic processes that take place during exercise [27–29]. Additionally, intracellular solute concentrations might be elevated during exercise as a result of phosphocreatine splitting and increased lactate and H^+, ensuing water influx in order to retain osmotic balance [28] and could contribute to myofiber swelling as well. This volume increase is followed by a compensatory mechanism named the regulatory volume decrease (RVD) that releases electrolytes (e.g., K^+, HCO_3^-, Cl^-) and osmolytes such as taurine concurrently with water in order to normalize cellular volume [19,20,26–31]. Thus, taurine is released in order to counteract myofiber swelling, a phenomenon which occurs during exercise [28,31].

2.3. Taurine and Its Role in Protein and Membrane Stabilization

The stabilizing effect of taurine is mentioned in many papers. However, the mechanism by which taurine is able to exert stabilization is poorly described. In order to comprehend this characteristic, it is important to understand its interaction with water molecules, considering the chemical properties related to its molecular structure. One of the most popular hypotheses that could explain protein stabilization by osmolytes is based on preferential exclusion [32–35]. This principle builds on unfavorable interactions between proteins and osmolytes in terms of Gibbs adsorption isotherm [32]. In a denatured state, the area of the peptide backbone by which osmolytes can interact is larger and results in increased Gibbs energy (unfavorable). In order to reduce these interactions, the thermodynamic component drives the folding equilibrium towards its native state, also referred to as the osmophobic effect, which is associated with a much lower Gibbs energy. This simplified explanation implies that in the presence of stabilizing osmolytes, the Gibbs energy of the denatured state is much higher compared to Gibbs energy of the folded state [32]. Therefore, the folded protein conformation is favored and osmolytes act as protein stabilizers [32,33]. In general, the presence of osmolytes results in a specific distribution of water molecules around the proteins in a preferential hydrated state and osmolyte exclusion from the protein backbone [32–35].

Furthermore, stabilizing actions have been attributed to taurine, as well as direct interaction with protein side chains [33–35]. The amino group of taurine orients itself preferentially to the protein side chain. This strengthens the hydrogen bonded network of water surrounding the protein and stabilizes its native form. The latter appears to contradict the preferential exclusion theory; however, such interactions between osmolytes and side chains have also been discussed by Bolen et al. [32]. It should be noted that protein side chains are associated with other characteristics than the protein backbone and favorable interactions between osmolyte and side chains might occur. Supposedly, the latter does not substantially alter the protein folding state [32]. In the article by Brudziak et al., the protein was hydrated in the presence of taurine [35]. This might suggest that besides limited interactions between osmolytes and protein side chains, the protein is still preferentially surrounded by water molecules. Taurine was able to increase the thermal stability of both lysozyme and ubiquitin protein [35–38], although the extent of stabilization was protein specific [35].

In addition to protein stabilization, membrane stabilizing properties of taurine were hypothesized by Huxtable and Bressler [39]. Taurine inhibits the activity of phospholipid methyltransferase, which catalyzes the methylation of phosphatidylethanolamine to form phosphatidylcholine and thus taurine could alter the composition and consequently the properties and stability of phospholipid membranes [40–42]. In evidence, the presence of taurine decreased the viscosity of erythrocytic membranes, suggesting taurine might increase membrane fluidity [43].

Eccentric muscle contraction might induce denaturation of myofibrillar proteins as hypothesized by Paulsen et al. [44]. In addition, the unfolded protein response (UPR) is activated during exercise [45,46] which might indicate that proteins struggle to maintain native folding conformations. Interestingly, prolonged exercise increased the denaturation temperature of albumin, pointing to enhanced thermal stability [47]. The importance of taurine in protein stabilization is illustrated in the TauT KO mouse model [15]. It is assumed that a lack of taurine allows accumulation of unfolded and/or misfolded proteins in skeletal muscle which activates expression of genes involved in the UPR. Thus, taurine plays a key role in protein homeostasis of skeletal muscle [15,48].

2.4. Taurine and Its Role in Oxidative Stress

Under physiological conditions, reactive oxygen species (ROS) are balanced by antioxidant mechanisms that detoxify reactive species. A limited amount of ROS is produced during exercise and exerts advantageous effects on force generation. In addition, low levels of ROS might protect against injury through adaptations in cellular signaling upon regular training exercise [27,47,49], whereas high levels of ROS are associated with muscle dysfunction [50]. Although direct scavenging of the main ROS (e.g., superoxide anion (O_{2-}), hydrogen peroxide (H_2O_2), and hydroxyl radical ($\cdot$OH)) by taurine is considered unlikely [51,52], taurine is believed to relieve oxidative stress through neutralization of hypochlorous acid (HOCl) and upregulation of antioxidant enzymes [53–55]. Upon inflammation, neutrophils become activated and release myeloperoxidase (MPO). MPO catalyzes the reaction of chloride and H_2O_2, a classic ROS molecule, resulting in the formation of HOCl. HOCl has toxic effects and interferes with cellular processes including molecular transport and pump capacity [56]. Recently, it has been hypothesized that HOCl could alter excitation-contraction (E-C) coupling and impairs muscle force production [57]. HOCl is converted to TauCl after interaction with taurine. TauCl possesses anti-inflammatory properties and increases antioxidant enzymes including heme oxygenase 1 in murine microglial cells and muscle cells [58,59]. Taurine supplementation increased activity of antioxidant enzymes such as superoxide dismutase and catalase, measured in the blood of patients with type II diabetes [60]. Similar results were obtained in the liver and kidney of an ethanol-induced oxidative stress mouse model [61].

Furthermore, antioxidant effects of taurine on superoxide production and lipid peroxidation were observed in the muscle of eccentric exercised rats [62]. Similarly oxidative lipid damage was reduced by taurine treatment in a mouse model of muscle overuse [63].

2.5. The Role of Taurine in Mitochondrial Protein Synthesis

Taurine participates in the synthesis of mitochondrial proteins, more specifically proteins that require $tRNA^{(Leu)}$ and $tRNA^{(Lys)}$ with uridine on a Wobble position [64]. If the first base of the tRNA anticodon is uridine, then the classic Watson–Crick rules are substituted by Wobble base pairing rules. According to this hypothesis, H-bonds are formed between the first base of the anticodon (tRNA) and the third position of the mRNA codon. Contrastingly to Watson–Crick base pairing, Wobble pairing suggests that uridine on a Wobble position at the anticodon can form H-bonds with A, U, G, and C on the third position of the mRNA codon mRNA. Whereas, taurine-conjugated uridine tRNA will promote the formation of H-bonds between uridine and A or G on the third codon position and is required for appropriate translation to leucine (UUA/UUG) [64]. Taurine modification of uridine is required in some mitochondrial tRNAs to ensure a more specific codon-anticodon interaction and proper mitochondrial protein synthesis. Cytochrome b, ND5, and ND6 are mitochondrial proteins containing UUG codons, and synthesis of these proteins is potentially hampered if taurine conjugation of mitochondrial $tRNA^{Leu(UUR)}$ is deficient. ND5 and ND6 are functional subunits of oxidative phosphorylation complex I, that catalyzes electron transport from NADH to coenzyme Q [64,65]. As the process of oxidative phosphorylation is one of the main sources of ROS [50] in myofibers, preservation of mitochondrial function is considered essential in the safeguarding of oxidative stress.

Specific mutations in $tRNA^{(Lys)}$ and $tRNA^{(Leu)}$ are associated with respectively myoclonic epilepsy with ragged red fibers (MERRF) and mitochondrial encephalopathy, lactic acidosis, and stroke-like episodes (MELAS). MERRF and MELAS are mitochondrial diseases that show disturbed protein synthesis and pathological characteristics such as exercise intolerance, thus of which some aspects resemble the TauT KO phenotype [9,64,66]. Of note, taurine supplementation in MELAS patients improved the occurrence of stroke-like episodes [67].

2.6. Taurine and Its Role in Calcium Homeostasis

Calcium is an essential cellular building block and a universal carrier of biological information. As a diffusible intracellular second messenger, calcium is involved in signaling transduction pathways and can regulate many different processes ranging from neurotransmitter release, bone formation, and blood coagulation to muscle contraction [68]. More specifically, calcium is essential during the E-C coupling mechanism that transforms the electrical input (action potential) to a mechanical output (contraction). Upon adequate stimulation, an action potential is generated and propagates over the sarcolemma to the transverse tubules that contain L-type calcium channels e.g., dihydropyridine receptors (DHPR). In the skeletal muscle, DHPR are in close contact with the ryanodine receptors (RyR) of the sarcoplasmic reticulum (SR), which releases calcium into the cytosol. Binding of calcium to troponin C induces the translocation of tropomyosin, thereby allowing the formation of cross-bridges between actin and myosin filaments and subsequent contraction [69].

Temporal and spatial changes of calcium concentration in cytoplasm or organelles are monitored by a multitude of calcium sensing proteins that determine the character and duration of the cellular response. The SR is an organelle involved in storage of calcium which is released into the cytoplasmic environment upon muscle stimulation. Taurine partially preserved SR function upon exposure to phospholipase C, which is known to hamper calcium transport, in SR derived from rat skeletal muscle. Furthermore, in the presence of 15 mM taurine, uptake of calcium oxalate by the SR was increased by more than 20% [39].

Similarly, taurine significantly increased the accumulation of calcium in SR of skinned EDL rat myofibers [70]. In this experiment, the membrane of the myofibers was mechanically removed which allowed to control intracellular taurine concentrations. The enhanced calcium SR load and subsequent release upon stimulation might explain the increment in force response upon depolarization. The authors hypothesized that taurine modulates SR calcium pump activity, allowing increased calcium accumulation [70]. These results are consistent with observations in human skeletal muscle fibers type I and II; interestingly, these different muscle fibers contain a different type of SR calcium transport ATPase (SERCA)-isoform, thus taurine might modulate either activity and/or calcium affinity [71]. Of note, a small decline in calcium sensitivity was observed in the presence of taurine in SR of skinned EDL myofibers [70].

Furthermore, voltage clamp experiments carried out in cardiomyocyte derived from guinea pig showed that the effect of taurine on calcium influx is highly dependent of the extracellular calcium concentration. In the presence of taurine (20 mM), calcium influx was slightly increased upon low extracellular calcium concentration (0.8 mM) and vice versa, the calcium influx was slightly yet significant decreased when extracellular calcium concentrations were high (3.6 mM). No effect of taurine on inward calcium current could be observed at an extracellular calcium level of 1.8 mM, suggesting that taurine aims at maintaining intracellular calcium homeostasis [72]. However, taurine modulated resting membrane potential regardless of extracellular calcium channels.

3. Pathophysiological Characteristics of Taurine in DMD

3.1. Regulation in Dystrophin Deficiency

Previously, it has been suggested that alterations in taurine and/or its regulation associate with dystrophinopathology [73–75]. Lack of dystrophin results in destabilization of DAPC and thereby rendering the sarcolemma more susceptible to contraction-induced damage. Membrane fragility is observed in dystrophin deficient muscle as evidenced by increased permeability to dyes such as Evans Blue and Procion Orange, accumulation of serum proteins such as albumin and IgG, increased levels of creatine kinase in the blood, and sarcolemma rupture in myofibers [7,76,77].

Taurine regulation is altered in dystrophic tissues of mouse models. Some studies reported a significant downregulation of the TauT in muscles of the mdx mouse model [73], whereas others reported no significant differences between mdx mice and age-matched control mice [78,79]. Seemingly, TauT is significantly downregulated in young mdx mouse and normalizes to control levels over time.

Similar to the expression of TauT, differences in muscle taurine content with age were reported as well, yet with considerable differences between studies. Taurine content of mdx muscle was comparable to controls at age 3–4 weeks, but increased by age 10 weeks [78,80]. A significant downregulation of taurine in mdx mice aged 4 and 6 weeks compared to age-matched control mice was reported [59,73], whereas other studies found no significant difference in taurine content between control and mdx mice at age 6 weeks [73,81]. A decline in taurine content was on the other hand observed in the plantaris muscle of 6-month-old mdx mice compared to wet weight yet did not hold up when compared to dry weight [79]. Contrasting results have been reported regarding the taurine content in mdx mice, however most of these studies conclude that taurine is differentially regulated in mdx mice compared to control mice at a certain time point. The expression of TauT and taurine content in the muscle of mdx mice compared to age-matched control mice is summarized in Table 1. Interestingly, muscle taurine levels increased in glucocorticoid treated mdx mice, pointing to its reestablishment by immunosuppressive therapeutic interventions [75].

Table 1. Summary of TauT and taurine expression in the mdx mouse.

Timepoint of Analysis	Muscle TauT Content	Muscle Taurine Content	Muscle Type	Reference
18 days	↓ in mdx mice	≈ controls	quadriceps	73
22 days	/	≈ controls	quadriceps	80
28 days	≈ controls	↓ in mdx mice	quadriceps	73
28 days	≈ controls	≈ controls	tibialis anterior	78
6 weeks	↓ in mdx mice	≈ controls	quadriceps	73
6 weeks	/	↓ in mdx mice	quad/gas	59
6 weeks	/	≈ controls	tibialis anterior	81
10 weeks	≈ controls	↑ in mdx	tibialis anterior	78
6 months	≈ controls	↓ in mdx mice (wet weight) ≈ controls (dry weight)	EDL (TauT); plantaris (taurine)	79

As opposed to the observations in the mdx mouse model a significant increase in muscle taurine and its transporter were discerned in 8-month-old Golden Retriever Muscular dystrophy (GRMD) canine model [74]. Similarly, taurine regulation in DMD patients differs from healthy control patients, as shown by increased levels of muscle TauT protein in DMD patients [82,83]. Thus, expression of TauT is differently regulated in the mdx mouse than in the GRMD-model and DMD patients. Interestingly, the phenotype in the mdx mouse model is milder than in GRMD dogs and DMD patients. Myofiber necrosis and muscle weakness become conspicuous approximately 3 weeks after mdx mice are born. Myofiber necrosis in the mdx mouse is most explicit in the young to juvenile period, followed by necrosis at a slower pace in adult mice [74,80]. The course of disease in mdx mice thereby differs from the more progressive pathology in GRMD dogs and DMD patients, that is characterized by fatty replacement and fibrosis at an early age [74]. These differences in pathological features might be linked to regulation of TauT.

3.2. Role in Oxidative Stress Management and Mitochondrial Protein Synthesis

Mitochondria participate in cellular energy production by synthesizing adenosine triphosphate (ATP) through oxidative phosphorylation [84]. This process includes transfer of electrons, required for ATP-synthesis, and is inevitably linked to production of ROS. Under physiological conditions, basal levels of ROS are generated as a by-product of oxidative phosphorylation. However, ROS production is enhanced upon mitochondrial dysfunction [84]. Upon oxidative phosphorylation, electron transfer is facilitated by mitochondrial protein complexes that resides within the inner mitochondrial membrane [84]. However, Onopiuk found significantly reduced protein levels of complex III, cytochrome-c reductase, and complex V, the ATP synthase, in immortalized myoblasts of mdx mice. Of note, myoblasts of mdx mice (SC-5), and control myoblasts (IMO) were used [85]. Neither of these myoblasts expressed dystrophin protein, only dystrophin mRNA was present in control myoblasts. In addition, Onopiuk et al. showed an increase in mitochondrial membrane potential [85]. Taken together, these results suggest mitochondrial dysfunction and inevitably, increased ROS production in dystrophin deficient cells. Taurine reduces ROS by upregulation of antioxidant enzymes and might preserve electron transport chain activity by safeguarding mitochondrial protein synthesis of subunits involved in the respiratory chain. In evidence, reduced taurine levels hamper expression of ND6, subunit of complex I in the mitochondria of cardiomyocytes derived from rat [64,86].

3.3. Dysregulation of Calcium Homeostasis

Excessive calcium levels are observed in dystrophic myofibers; however, calcium entry through sarcolemmal tears is not considered as the main contributor to calcium overload in dystrophic myofibers. Apparently, the open probability of calcium leak channels in the proximity of micro-tears are increased, thereby allowing an increased calcium influx [87].

Additionally, increased expression and activation of store-operated calcium channels, presumably induced by calcium-independent phospholipase A2, is observed in dystrophin deficient myofibers and could contribute to calcium overload as well [84,88–90]. Moreover, a correlation between the dystrophic phenotype and expression of a stretch-activated channel, transient receptor potential canonical (TRPC) channel 1, was discovered in different muscles of the mdx mouse. The diaphragm of mdx mice was affected the most, as shown by increased Evans Blue permeability, and showed a significant upregulation of TRPC1 expression compared to controls [88]. Thus, the involvement of the TRPC channels might also contribute to the calcium overload as evidenced by increased expression of various TRPC channels in mdx mice. The cytoplasmic calcium concentration is not only determined by calcium channels/exchangers but also by the SR uptake and release of calcium, which plays an essential role during E-C coupling. RyR, responsible for the release of calcium from SR upon depolarization, is hyper nitrosylated in dystrophic muscle and consecutive depletion of calstabin-1 upon RyR-nitrosylation results in calcium leakage [84,89–92]. During relaxation, cytoplasmic calcium ions are sequestered by SERCA, thereby lowering its cytoplasmic concentration. In the mdx mouse model, the removal of calcium ions by the SR is hampered, suggesting reduced SERCA functioning. In conclusion, calcium homeostasis is disturbed in dystrophic muscle on many levels. Chronic cytoplasmic calcium overload will induce activation of degrading pathways mediated by phospholipase 2 and proteases which eventually can lead to myofiber death [7,90].

In addition, cytoplasmic calcium overload can induce accumulation of calcium in the mitochondria and mitochondrial dysfunction. Multiple pathways have been proposed by which mitochondrial calcium overload can induce ROS production, such as mitochondrial permeability transition (MPT)-mediated release of anti-oxidative enzymes, dislocation of mitochondrial proteins involved in electron transports including cytochrome C, induction of NO, etc. [84,93]. Mitochondrial calcium overload can elicit MPT pore formation. A permeable pore is formed that spans inner and outer mitochondrial membranes and results in mitochondrial swelling [84]. Furthermore, dystrophic muscle cells show an increased susceptibility to calcium and therefore are more prone to MPT pore formation, that eventually cause mitochondrial swelling and death [87,91]. In evidence, mitochondrial swelling, indicative for MPT pore formation, was induced at a lower calcium load in mitochondria compared to WT [84,90,92–94]. Interestingly, taurine is able to attenuate calcium-induced swelling of mitochondria derived from skeletal muscle [95].

4. Taurine Supplementation as a Therapeutic Strategy for DMD

4.1. Effect of Taurine on Muscle Force

A beneficial effect of taurine on muscle force remains controversial: supported by some studies yet disproved by others. A significant increase in peak twitch force was obtained in taurine supplemented (2.5% *w*/*v* in drinking water) mdx mice compared to untreated mdx mice at 4 weeks. However, this effect was abrogated in juvenile and adult mdx mice, respectively aged 10 weeks (2.5% *w*/*v*) and 6 months (3% *w*/*v*) [78,79]. Similarly, a >50% increase in maximum isometric tetanic specific force (sPo) was obtained by taurine supplementation in mdx mice aged 4 and 6 weeks [78,81]. However, no treatment effect (1 g/kg/day) on specific tetanic force was detected in muscle of mice aged 5–7 weeks, whereas fore limb force, assessed by means of a grip strength meter, was ameliorated [96]. Furthermore, a study performed by Barker reported no effects of taurine treatment (2.5% *w*/*v*) on peak twitch force, maximum specific force, nor fatigue or recovery in mdx mice treated from week 2 until week 4 [97]. Similarly, no effect on maximum specific force in 6-week-old mdx mice was observed when high doses of taurine (16 g/kg/day) were administered [98].

In 6-month-old mdx mice, taurine could not ameliorate specific maximum isometric force production. However, after a fatigue protocol consisting of recurrent electrical stimulation, the EDL of taurine treated mdx animals was more resistant to fatigue, as shown by a significant higher force production at the end of stimulation relative to force production at the beginning. Furthermore, EDL muscle of taurine treated mdx animals showed a significantly better recovery capacity at 10–60 min after stimulation than untreated mdx animals [79]. Taurine treatment in exercised mdx mice significantly increased in vivo forelimb muscle strength normalized to body weight [96,99], but did not alter locomotor activity of mdx mice. A similar finding was reported in a study that supplemented unexercised mdx mice. Taurine supplementation (±4 g/kg/day) in 6-week-old unexercised mdx mice significantly increased grip strength [81]. Whereas no effect of high taurine treatment (16 g/kg/day) was observed on normalized grip strength [98].

In summary, benefit of taurine supplementation on muscle force differed considerably between published studies. Although multiple studies have been executed, these used different concentrations, ingestion methods, and treatment protocols, which might explain the obtained conflicting results, and which hampers interpretation.

4.2. *Effect of Taurine on Oxidative Stress and Inflammation*

Although different treatment conditions and read-outs were used, multiple studies have shown anti-inflammatory and anti-oxidative effects of taurine treatment in mdx mice [59,80,81,95]. The anti-oxidant and anti-inflammatory actions of taurine have been proposed as a mechanism by which taurine protects dystrophic tissue from damage. In evidence, muscle of taurine-treated mdx mice showed significantly less NF-κβ positive fibers, TNF-α levels, neutrophil elastase, and MPO activity [59,80,81,96]. Accordingly, taurine-treated mdx mice showed lower levels of disulfide and protein thiol oxidation in muscle compared to untreated mdx mice, which could point towards protection against oxidative stress [59,81,98,100]. Similarly, ROS levels were significantly reduced upon taurine treatment in muscles of exercised dystrophic mice, as shown by dihydroethidium (DHE) staining, which reacts with O_{2^-} [96]. In addition, taurine normalized the resting macroscopic ionic conductance (gm), a measure for ROS production, to WT levels [96].

4.3. *Effects of Taurine on E-C Coupling*

In mdx mice, the threshold of the membrane potential at which a contraction is elicited is shifted towards a more negative value than in control mice. Thus, contraction is induced upon a lower depolarization state than in WT animals [96,101]. This mechanical threshold is indicative for E-C functioning and suggests alterations in E-C-coupling and/or calcium homeostasis [96,101]. Interestingly, taurine treatment in mdx mice has been shown to partially restore the mechanical threshold [96,99,101]. Taurine treatment in mdx did not alter expression of proteins involved in E-C coupling such as calcium sensitive receptors (RyR), calcium channels (DHPR channels), calcium ATP-ase pump (SERCA) and calcium binding proteins (calsequestrin) [78]. Contrastingly, taurine supplementation (2.5% *w*/*v*) in rats significantly increased expression of calsequestrin 1 in the muscle [102]. Although in the mdx mouse and rat study the same dose of taurine was used (2.5% *w*/*v*), this discrepancy in outcome might be explained by age-dependent calsequestrin regulation, changes in calsequestrin regulation upon dystrophin deficiency, species-dependent regulation, or differences in treatment protocol. However, a specific explanation cannot be pinpointed with the current literature studies available.

4.4. *Effect of Taurine on Histopathological Characteristics of the Mdx Mouse*

Histopathological characteristics such as necrotic myofibers and fibers with centralized nuclei, indicative for regeneration ensuing myofiber damage, are conspicuous in Hematoxylin-Eosin stained sections of mdx muscles and are used to evaluate therapeutic effectiveness [59]. In the study of Barker, taurine (2.5% *w*/*v*) reduced the amount of non-contractile area in young mdx mice, whereas no effect on the percentage regenerative fibers could be observed [78]. Other studies have reported a significant increase in healthy myofibers with peripheral nuclei upon taurine treatment [59], a decrease of histopathological features and myofiber necrosis [80,96]. Thus, taurine seems to alleviate histological features related to dystrophinopathy.

4.5. Combinatory Use of Taurine and Glucocorticoids

Combined use of taurine (1 g/kg) and α-methylprednisolone (PDN) (1 mg/kg), a synthetic adrenocortical hormone, in the exercised mdx mouse model significantly improved muscle strength in such a way that a synergistic effect was proposed by Cozzoli et al. [103]. The increase in fore limb muscle strength after 4 weeks of treatment was significantly higher in mice that received combined therapy compared to mice treated with PDN. Similarly, the increment in muscle force normalized to body weight was remarkably elevated in mice that received taurine + PDN compared to single-drug treatments. Furthermore, combined treatment normalized the rheobase potential to WT values, however similar results were observed in taurine-treated mice. No synergistic effect of combination treatment compared to PDN-treatment was detected regarding histopathological markers that included the amount of centronucleated fibers, necrosis and non-muscle tissue [103].

Contrastingly, the study of Barker et al. reported no effects of combined treatment on muscle strength. These opposing findings might be explained by differences in experimental set-up since in the latter study therapeutic intervention was initiated more closely to onset of damage, higher taurine concentrations were used for a shorter period of time and analysis occurred at the peak of damage [97]. Besides possible synergistic effects of combined treatment, taurine could also counteract side-effects related to corticosteroids. Dexamethasone causes muscle atrophy and significantly lowers the myotube diameter, which is restored upon taurine treatment [18]. Furthermore, taurine protects against glucocorticoid-induced mitochondrial dysfunction of the bone and might attenuate corticoid-induced osteoporosis and or osteonecrosis, pathological features that are conspicuous in glucocorticoid-treated DMD patients [104–106].

4.6. Potential Caveats of Taurine Treatment

In some rat studies, different test regimes of taurine supplementation (3% w/v for 4 weeks [107], 100 mg/kg for 2 weeks [108]; 500 mg/kg for 2 weeks [109]) resulted in increased muscle taurine content, whilst other long-term studies (1% $w/v \approx$ 50 mg/kg for 22 weeks [110]) could not report increased muscle taurine levels. Similarly, in mouse studies muscle taurine levels were elevated upon taurine supplementation (4% w/v for 3 weeks [59]; 3% w/v for 4 weeks [79]). Interestingly, one study showed that continuous taurine supplementation (2.5% w/v) in mdx mice increased muscle taurine concentration up to the age of 4 weeks, but when treatment was prolonged up to the age of 10 weeks, muscle taurine concentration was significantly decreased in taurine supplemented mdx mice compared to untreated mdx mice [78]. Similarly, high doses of taurine (8% $w/v \approx$ 16 g/kg/day) added to the drinking water up until the age of 6 weeks, did not increase muscle taurine content [98]. Therefore, it might be hypothesized that the muscle taurine content is strictly regulated. Since chronic treatment might not be able to increase intramuscular taurine levels it is not known if long-term taurine treatment could effectively attenuate muscle damage. Furthermore, taurine intake (≈5 g/day) for a period of 7 days was reported not to alter muscle taurine levels in humans [111].

In general, taurine is well tolerated and safe if used in appropriate concentrations [112]. One study has reported gastro-intestinal complaints; however, taurine was used in combination with other nutritional supplements [112,113]. Another study aimed to investigate taurine supplementation in patients with end-stage renal failure. In healthy subjects, excessive taurine is excreted; however, due to kidney failure the taurine overload was not immediately cleared and these patients suffered from dizziness and vertigo [114] which was the reason to discontinue the study.

High taurine treatment (16 g/kg) in 6 week-old animals significantly reduced the body weight of mdx mice by more than 20% and shortened tibia length by 10%. Of note, the body weight and tibia length of untreated mdx mice were comparable to those of WT animals in this experiment [98]. Similarly, taurine treatment (3% w/v) for 4 weeks in adult mdx mice (5 months old) reduced body weight and muscle mass of mdx mice; however, the weight of these animals still exceeded that of WT animals [79]. Interestingly, no effect of taurine on body weight was observed in WT animals. Obesity is commonly observed in DMD patients, especially in glucocorticoid-treated patients [7,115,116]. Therefore, it remains unclear if taurine-induced decreased body weight would be disadvantageous in these patients. Obviously if growth is hampered, this should be avoided. A study conducted in piglets reported a reduced gain to feed ratio upon higher taurine supplementation (3% taurine diet) compared to untreated piglets whilst supplementation with 0.3% taurine might improve growth [117]. Terrill et al. have hypothesized that high taurine supplementation in young animals interferes with the taurine synthesis pathway, shifting the equilibrium to the left inducing increased cysteine levels in the plasma which could hamper growth [98].

4.7. Other Nutritional Supplements Used in Duchenne Muscular Dystrophy

Approximately 50–65% of DMD patients are using vitamins or nutritional supplements [115,116], which indicates that the quest for supportive treatment, including but not limited to taurine, in DMD is relevant. A detailed overview of nutritional supplements that are under investigation for the treatment of DMD is provided in the review of Boccanegra et al. [118].

The current nutritional guidelines for patients are very similar to those used for the general population. If serum 25-hydroxy-vitamin D drops below 30 ng/mL or calcium intake is low, the use of vitamin D and, respectively, calcium intake under the form of calcium citrate or calcium carbonate is advised in order to stimulate bone health, which could be impaired in DMD patients [116,117,119]. In addition, other nutrients are currently under investigation. Coenzyme Q10, a naturally occurring anti-oxidant, significantly improved muscle strength in steroid-treated DMD patients [118,120]. Similarly, beneficial effects of L-arginine, creatine, omega-3 fatty acids have been observed in clinical trials [118].

5. Conclusions

Taurine is involved in numerous processes such as protein stabilization and osmotic homeostasis and appears to be indispensable for physiological muscle function as evidenced in the TauT KO mouse. We further zoomed in on DMD, a severe muscle disorder that showed altered regulation of taurine and its transporter. Multiple facets of dystrophinopathology and potential mechanisms on how taurine might act on these pathological features have been proposed and discussed in view of the mdx mouse model. Promising results have been obtained in the mdx mouse model in terms of inflammation, muscle strength, oxidative stress, etc. and led to the conclusion that taurine supplementation is relevant for DMD pathology. Former clinical trials conducted to evaluate the anti-aging or mood stabilizing effects of taurine have shown that taurine is well tolerated and considered safe upon appropriate use. However, up until now, no clinical trials have been conducted that evaluated taurine as a treatment for DMD patients. We propose that taurine has the potential to act as supportive therapy in combination with glucocorticoids for the treatment of DMD patients and further studies should be conducted to evaluate the effectiveness of chronic taurine supplementation.

Funding: This research received no external funding.

Conflicts of Interest: The authors declare no conflict of interest.

References

1. Wen, C.; Li, F.; Zhang, L.; Duan, Y.; Guo, Q.; Wang, W.; Yin, Y. Taurine is involved in energy metabolism in muscles, adipose tissue, and the liver. *Mol. Nutr. Food Res.* **2019**, *63*, 1800536. [CrossRef]
2. Warskulat, U.; Flögel, U.; Jacoby, C.; Hartwig, H.G.; Thewissen, M.; Merx, M.W.; Häussinger, D. Taurine transporter knockout depletes muscle taurine levels and results in severe skeletal muscle impairment but leaves cardiac function uncompromised. *FASEB J.* **2004**, *18*, 577–579. [CrossRef]
3. De Luca, A.; Pierno, S.; Camerino, D.C. Taurine: The appeal of a safe amino acid for skeletal muscle disorders. *J. Transl. Med.* **2015**, *13*, 243. [CrossRef] [PubMed]
4. Crisafulli, S.; Sultana, J.; Fontana, A.; Salvo, F.; Messina, S.; Trifirò, G. Global epidemiology of Duchenne Muscular Dystrophy: An updated systematic review and meta-analysis. *Orphanet J. Rare Dis.* **2020**, *15*, 141. [CrossRef] [PubMed]

5. Gloss, D.; Moxley, R.T.; Ashwal, S.; Oskoui, M. Practice guideline update summary: Corticosteroid treatment of Duchenne Muscular Dystrophy: Report of the guideline development subcommittee of the American academy of neurology. *Neurology* **2016**, *86*, 465–472. [CrossRef] [PubMed]
6. Quattrocelli, M.; Zelikovich, A.S.; Salamone, I.M.; Fischer, J.A.; McNally, E.M. Mechanisms and clinical applications of glucocorticoid steroids in muscular dystrophy. *J. Neuromuscul. Dis.* **2021**, *8*, 39–52. [CrossRef] [PubMed]
7. Duan, D.; Goemans, N.; Takeda, S.I.; Mercuri, E.; Aartsma-Rus, A. Duchenne Muscular Dystrophy. *Nat. Rev. Dis. Primers* **2021**, *7*, 1–19. [CrossRef] [PubMed]
8. Sun, C.; Shen, L.; Zhang, Z.; Xie, X. Therapeutic strategies for Duchenne Muscular Dystrophy: An update. *Genes* **2020**, *11*, 837. [CrossRef] [PubMed]
9. Ito, T.; Yoshikawa, N.; Schaffer, S.W.; Azuma, J. Tissue taurine depletion alters metabolic response to exercise and reduces running capacity in mice. *Amino Acids* **2014**, *2014*, 964680. [CrossRef] [PubMed]
10. Warskulat, U.; Heller-Stilb, B.; Oermann, E.; Zilles, K.; Haas, H.; Lang, F.; Häussinger, D. Phenotype of the taurine transporter knockout mouse. *Meth. Enzymol.* **2007**, *428*, 439–458.
11. Ito, T.; Kimura, Y.; Uozumi, Y.; Takai, M.; Muraoka, S.; Matsuda, T.; Azuma, J. Taurine depletion caused by knocking out the taurine transporter gene leads to cardiomyopathy with cardiac atrophy. *J. Mol. Cell. Cardiol.* **2008**, *44*, 927–937. [CrossRef] [PubMed]
12. De Paepe, B. Osmolytes as mediators of the muscle tissue's responses to inflammation: Emerging regulators of myositis with therapeutic potential. *Eur. Med. J. Rheumatol.* **2017**, *4*, 83–89.
13. Ito, T.; Oishi, S.; Takai, M.; Kimura, Y.; Uozumi, Y.; Fujio, Y.; Azuma, J. Cardiac and skeletal muscle abnormality in taurine transporter-knockout mice. *J. Biomed. Sci.* **2010**, *17*, S20. [CrossRef] [PubMed]
14. Hamilton, E.J.; Berg, H.M.; Easton, C.J.; Bakker, A.J. The effect of taurine depletion on the contractile properties and fatigue in fast-twitch skeletal muscle of the mouse. *Amino Acids* **2006**, *31*, 273–278. [CrossRef]
15. Ito, T.; Yoshikawa, N.; Inui, T.; Miyazaki, N.; Schaffer, S.W.; Azuma, J. Tissue depletion of taurine accelerates skeletal muscle senescence and leads to early death in mice. *PLoS ONE* **2014**, *9*, e107409. [CrossRef]
16. Ito, T.; Miyazaki, N.; Schaffer, S.; Azuma, J. Potential anti-aging role of taurine via proper protein folding: A study from taurine transporter knockout mouse. *Adv. Exp. Med. Biol.* **2015**, *9*, 481–487.
17. Barbiera, A.; Sorrentino, S.; Lepore, E.; Carfì, A.; Sica, G.; Dobrowolny, G.; Scicchitano, B.M. Taurine attenuates catabolic processes related to the onset of sarcopenia. *Int. J. Mol. Sci.* **2020**, *21*, 8865. [CrossRef]
18. Uozumi, Y.; Ito, T.; Hoshino, Y.; Mohri, T.; Maeda, M.; Takahashi, K.; Azuma, J. Myogenic differentiation induces taurine transporter in association with taurine-mediated cytoprotection in skeletal muscles. *Biochem. J.* **2006**, *394*, 699–706. [CrossRef]
19. Brocker, C.; Thompson, D.C.; Vasiliou, V. The role of hyperosmotic stress in inflammation and disease. *Biomol. Concepts* **2012**, *3*, 345–364. [CrossRef]
20. Burg, M.B.; Ferraris, J.D.; Dmitrieva, N.I. Cellular response to hyperosmotic stresses. *Physiol. Rev.* **2007**, *87*, 1441–1474. [CrossRef]
21. Srinivas, S.P. Dynamic regulation of barrier integrity of the corneal endothelium. *Optom. Visc. Sci.* **2010**, *87*, E239–E254. [CrossRef]
22. Eveloff, J.L.; Warnock, D.G. Activation of ion transport systems during cell volume regulation. *Am. J. Physiol. Ren. Physiol.* **1978**, *252*, F1–F10. [CrossRef]
23. Lindinger, M.; Hawke, T.; Lipskie, S.; Schaefer, H.; Vickery, L. K+ transport and volume regulatory response by NKCC in resting rat hindlimb skeletal muscle. *Cell. Physiol. Biochem.* **2002**, *12*, 279–292. [CrossRef] [PubMed]
24. Baliou, S.; Kyriakopoulos, A.M.; Goulielmaki, M.; Panayiotidis, M.I.; Spandidos, D.A.; Zoumpourlis, V. Significance of taurine transporter (TauT) in homeostasis and its layers of regulation. *Mol. Med. Rep.* **2020**, *22*, 2163–2173. [CrossRef] [PubMed]
25. Ito, T.; Fujio, Y.; Hirata, M.; Takatani, T.; Matsuda, T.; Muraoka, S.; Azuma, J. Expression of taurine transporter is regulated through the TonE (tonicity-responsive element)/TonEBP (TonE-binding protein) pathway and contributes to cytoprotection in HepG2 cells. *Biochem. J.* **2004**, *382*, 177–182. [CrossRef]
26. Tsai, T.T.; Danielson, K.G.; Guttapalli, A.; Oguz, E.; Albert, T.J.; Shapiro, I.M.; Risbud, M.V. TonEBP/OREBP is a regulator of nucleus pulposus cell function and survival in the intervertebral disc. *J. Biol. Chem.* **2006**, *281*, 25416–25424. [CrossRef]
27. King, M.A.; Baker, L.B. Dehydration and exercise-induced muscle damage: Implications for recovery. *Sports Sci. Exch.* **2020**, *29*, 1.
28. King, M.A.; Clanton, T.L.; Laitano, O. Hyperthermia, dehydration, and osmotic stress: Unconventional sources of exercise-induced reactive oxygen species. *Am. J. Physiol. Regul. Integr. Comp. Physiol.* **2016**, *310*, R105–R114. [CrossRef] [PubMed]
29. Pivarnik, J.M.; Leeds, E.M.; Wilkerson, J.E. Effects of endurance exercise on metabolic water production and plasma volume. *J. Appl. Physiol.* **1984**, *56*, 613–618. [CrossRef] [PubMed]
30. Hoffmann, E.K.; Lambert, I.H.; Pedersen, S.F. Physiology of cell volume regulation in vertebrates. *Physiol. Rev.* **2009**, *89*, 193–277. [CrossRef]
31. Ørtenblad, N.; Young, J.F.; Oksbjerg, N.; Nielsen, J.H.; Lambert, I.H. Reactive oxygen species are important mediators of taurine release from skeletal muscle cells. *Am. J. Physiol. Cell Physiol.* **2003**, *284*, C1362–C1373. [CrossRef]
32. Bolen, D.W.; Baskakov, I.V. The osmophobic effect: Natural selection of a thermodynamic force in protein folding. *J. Mol. Biol.* **2001**, *310*, 955–963. [CrossRef]
33. Bruździak, P.; Adamczak, B.; Kaczkowska, E.; Czub, J.; Stangret, J. Are stabilizing osmolytes preferentially excluded from the protein surface? FTIR and MD studies. *Phys. Chem. Chem. Phys.* **2015**, *17*, 23155–23164. [CrossRef]

34. Bruzdziak, P.; Panuszko, A.; Stangret, J. Influence of osmolytes on protein and water structure: A step to understanding the mechanism of protein stabilization. *J. Phys. Chem.* **2013**, *117*, 11502–11508. [CrossRef] [PubMed]
35. Bruździak, P.; Panuszko, A.; Kaczkowska, E.; Piotrowski, B.; Daghir, A.; Demkowicz, S.; Stangret, J. Taurine as a water structure breaker and protein stabilizer. *Amino Acids* **2018**, *50*, 125–140. [CrossRef] [PubMed]
36. Arakawa, T.; Timasheff, S. The stabilization of proteins by osmolytes. *Biophys. J.* **1985**, *47*, 411–414. [CrossRef]
37. Abe, Y.; Ohkuri, T.; Yoshitomi, S.; Murakami, S.; Ueda, T. Role of the osmolyte taurine on the folding of a model protein, hen egg white lysozyme, under a crowding condition. *Amino Acids* **2015**, *47*, 909–915. [CrossRef]
38. Miyawaki, O.; Dozen, M.; Nomura, K. Thermodynamic analysis of osmolyte effect on thermal stability of ribonuclease A in terms of water activity. *Biophys. Chem.* **2014**, *185*, 19–24. [CrossRef] [PubMed]
39. Huxtable, R.; Bressler, R. Effect of taurine on a muscle intracellular membrane. *Biochim. Biophys. Acta Biomembr.* **1973**, *323*, 573–583. [CrossRef]
40. Punna, S.; Ballard, C.; Hamaguchi, T.; Azuma, J.; Schaffer, S. Effect of taurine and methionine on sarcoplasmic reticular Ca^{2+} transport and phospholipid methyltransferase activity. *J. Cardiovasc. Pharmacol.* **1994**, *24*, 286–292. [CrossRef]
41. Hamaguchi, T.; Azuma, J.; Schaffer, S. Interaction of taurine with methionine: Inhibition of myocardial phospholipid methyltransferase. *J. Cardiovasc. Pharmacol.* **1991**, *18*, 224–230. [CrossRef] [PubMed]
42. Lambert, I.H.; Kristensen, D.M.; Holm, J.B.; Mortensen, O.H. Physiological role of taurine–from organism to organelle. *Acta Physiolog.* **2015**, *213*, 191–212. [CrossRef] [PubMed]
43. Akhalaya, M.Y.; Kushnareva, E.A.; Parshina, E.Y.; Platonov, A.G.; Graevskaya, E.E. Membrane-modifying effect of taurine. *Biophysics* **2012**, *57*, 485–490. [CrossRef]
44. Paulsen, G.; Vissing, K.; Kalhovde, J.M.; Ugelstad, I.; Bayer, M.L.; Kadi, F.; Raastad, T. Maximal eccentric exercise induces a rapid accumulation of small heat shock proteins on myofibrils and a delayed HSP70 response in humans. *Am. J. Physiol. Regul. Integr. Comp. Physiol.* **2007**, *293*, R844–R853. [CrossRef]
45. Wu, J.; Ruas, J.L.; Estall, J.L.; Rasbach, K.A.; Choi, J.H.; Ye, L.; Spiegelman, B.M. The unfolded protein response mediates adaptation to exercise in skeletal muscle through a PGC-1α/ATF6α complex. *Cell Metab.* **2011**, *13*, 160–169. [CrossRef]
46. Estébanez, B.; de Paz, J.A.; Cuevas, M.J.; González-Gallego, J. Endoplasmic reticulum unfolded protein response, aging and exercise: An update. *Front. Physiol* **2018**, *9*, 1744. [CrossRef]
47. Mourtakos, S.; Philippou, A.; Papageorgiou, A.; Lembessis, P.; Zaharinova, S.; Hasanova, Y.; Koutsilieris, M. The effect of prolonged intense physical exercise of special forces volunteers on their plasma protein denaturation profile examined by differential scanning calorimetry. *J. Therm. Biol.* **2021**, *96*, 102860. [CrossRef]
48. Bhat, M.A.; Ahmad, K.; Khan, M.S.A.; Almatroudi, A.; Rahman, S.; Jan, A.T. Expedition into taurine biology: Structural insights and therapeutic perspective of taurine in neurodegenerative diseases. *Biomolecules* **2020**, *10*, 863. [CrossRef]
49. Steinbacher, P.; Eckl, P. Impact of oxidative stress on exercising skeletal muscle. *Biomolecules* **2015**, *5*, 356–377. [CrossRef]
50. Powers, S.K.; Ji, L.L.; Kavazis, A.N.; Jackson, M.J. Reactive oxygen species: Impact on skeletal muscle. *Compr. Physiol.* **2011**, *1*, 941.
51. Ábrigo, J.; Elorza, A.A.; Riedel, C.A.; Vilos, C.; Simon, F.; Cabrera, D.; Cabello-Verrugio, C. Role of oxidative stress as key regulator of muscle wasting during cachexia. *Oxid. Med. Cell. Long.* **2018**, *2018*, 2063179. [CrossRef] [PubMed]
52. Aruoma, O.I.; Halliwell, B.; Hoey, B.M.; Butler, J. The antioxidant action of taurine, hypotaurine and their metabolic precursors. *Biochem. J.* **1988**, *256*, 251–255. [CrossRef]
53. Gürer, H.; Özgünes, H.; Saygin, E.; Ercal, N. Antioxidant effect of taurine against lead-induced oxidative stress. *Arch. Environ. Contam. Toxicol.* **2001**, *41*, 397–402. [CrossRef]
54. Oudit, G.Y.; Trivieri, M.G.; Khaper, N.; Husain, T.; Wilson, G.J.; Liu, P.; Backx, P.H. Taurine supplementation reduces oxidative stress and improves cardiovascular function in an iron-overload murine model. *Circulation* **2004**, *109*, 1877–1885. [CrossRef] [PubMed]
55. Hagar, H.H. The protective effect of taurine against cyclosporine A-induced oxidative stress and hepatotoxicity in rats. *Toxicol. Lett.* **2004**, *151*, 335–343. [CrossRef]
56. Jaimes, E.A.; Sweeney, C.; Raij, L. Effects of the reactive oxygen species hydrogen peroxide and hypochlorite on endothelial nitric oxide production. *Hypertension* **2001**, *38*, 877–883. [CrossRef]
57. Lea, T.A.; Pinniger, G.J.; Arthur, P.G.; Bakker, T.J. Effects of HOCl oxidation on excitation–contraction coupling: Implications for the pathophysiology of Duchenne Muscular Dystrophy: Calcium signaling and excitation–contraction in cardiac, skeletal and smooth muscle. *J. Gen. Physiol.* **2021**, *154*, e2021ecc16. [CrossRef]
58. Seol, S.I.; Kim, H.J.; Choi, E.B.; Kang, I.S.; Lee, H.K.; Lee, J.K.; Kim, C. Taurine protects against postischemic brain injury via the antioxidant activity of taurine chloramine. *Antioxidants* **2021**, *10*, 372. [CrossRef] [PubMed]
59. Terrill, J.R.; Webb, S.M.; Arthur, P.G.; Hackett, M.J. Investigation of the effect of taurine supplementation on muscle taurine content in the mdx mouse model of Duchenne Muscular Dystrophy using chemically specific synchrotron imaging. *Analyst* **2020**, *145*, 7242–7251. [CrossRef]
60. Maleki, V.; Mahdavi, R.; Hajizadeh-Sharafabad, F.; Alizadeh, M. The effects of taurine supplementation on oxidative stress indices and inflammation biomarkers in patients with type 2 diabetes: A randomized, double-blind, placebo-controlled trial. *Diabetol. Metabol. Syndr.* **2020**, *12*, 9. [CrossRef]
61. Goc, Z.; Kapusta, E.; Formicki, G.; Martiniaková, M.; Omelka, R. Effect of taurine on ethanol-induced oxidative stress in mouse liver and kidney. *Chin. J. Physiol.* **2019**, *62*, 148. [CrossRef] [PubMed]

62. Silva, L.A.; Silveira, P.C.; Ronsani, M.M.; Souza, P.S.; Scheffer, D.; Vieira, L.C.; Pinho, R.A. Taurine supplementation decreases oxidative stress in skeletal muscle after eccentric exercise. *Cell Biochem. Funct.* **2011**, *29*, 43–49. [CrossRef]
63. Thirupathi, A.; Freitas, S.; Sorato, H.R.; Pedroso, G.S.; Effting, P.S.; Damiani, A.P.; Pinho, R.A. Modulatory effects of taurine on metabolic and oxidative stress parameters in a mice model of muscle overuse. *Nutrition* **2018**, *54*, 158–164. [CrossRef] [PubMed]
64. Jong, C.J.; Sandal, P.; Schaffer, S.W. The role of taurine in mitochondria health: More than just an antioxidant. *Molecules* **2021**, *26*, 4913. [CrossRef] [PubMed]
65. Leonard, J.V.; Schapira, A.H. Mitochondrial respiratory chain disorders I: Mitochondrial DNA defects. *Lancet* **2000**, *355*, 299–304. [CrossRef]
66. Schaffer, S.W.; Jong, C.J.; Ito, T.; Azuma, J. Role of taurine in the pathologies of MELAS and MERRF. *Amino Acids* **2014**, *46*, 47–56. [CrossRef] [PubMed]
67. Ohsawa, Y.; Hagiwara, H.; Nishimatsu, S.I.; Hirakawa, A.; Kamimura, N.; Ohtsubo, H.; Sunada, Y. Taurine supplementation for prevention of stroke-like episodes in MELAS: A multicentre, open-label, 52-week phase III trial. *J. Neurol. Neurosurg. Psychiatry* **2019**, *90*, 529–536. [CrossRef]
68. Manitshana, N. Calcium homeostasis. *S. Afr. J. Anaesth. Analg.* **2020**, *26*, S104–S107. [CrossRef]
69. MacIntosh, B.R.; Holash, R.J.; Renaud, J.M. Skeletal muscle fatigue–regulation of excitation–contraction coupling to avoid metabolic catastrophe. *J. Cell Sci.* **2012**, *125*, 2105–2114. [CrossRef] [PubMed]
70. Bakker, A.J.; Berg, H.M. Effect of taurine on sarcoplasmic reticulum function and force in skinned fast-twitch skeletal muscle fibres of the rat. *J. Physiol.* **2002**, *538*, 185–194. [CrossRef]
71. Dutka, T.L.; Lamboley, C.R.; Murphy, R.M.; Lamb, G.D. Acute effects of taurine on sarcoplasmic reticulum Ca^{2+} accumulation and contractility in human type I and type II skeletal muscle fibers. *J. Appl. Physiol.* **2014**, *117*, 797–805. [CrossRef]
72. Sawamura, A.; Sada, H.; Azuma, J.; Kishimoto, S.; Sperelakis, N. Taurine modulates ion influx through cardiac Ca^{2+} channels. *Cell Calcium* **1990**, *11*, 251–259. [CrossRef]
73. Terrill, J.R.; Grounds, M.D.; Arthur, P.G. Taurine deficiency, synthesis and transport in the mdx mouse model for Duchenne Muscular Dystrophy. *Int. J. Biochem. Cell Biol.* **2015**, *66*, 141–148. [CrossRef] [PubMed]
74. Terrill, J.R.; Duong, M.N.; Turner, R.; Le Guiner, C.; Boyatzis, A.; Kettle, A.J.; Arthur, P.G. Levels of inflammation and oxidative stress, and a role for taurine in dystropathology of the Golden Retriever muscular dystrophy dog model for Duchenne Muscular Dystrophy. *Redox Biol.* **2016**, *9*, 276–286. [CrossRef] [PubMed]
75. McIntosh, L.; Granberg, K.E.; Brière, K.M.; Anderson, J.E. Nuclear magnetic resonance spectroscopy study of muscle growth, *mdx* dystrophy and glucocorticoid treatments: Correlation with repair. *NMR Biomed.* **1998**, *11*, 1–10. [CrossRef]
76. Deconinck, N.; Dan, B. Pathophysiology of Duchenne Muscular Dystrophy: Current hypotheses. *Pediatr. Neurol.* **2007**, *36*, 1–7. [CrossRef] [PubMed]
77. Kornegay, J.N. The golden retriever model of Duchenne Muscular Dystrophy. *Skelet. Muscle* **2017**, *7*, 9. [CrossRef]
78. Barker, R.G.; Horvath, D.; van der Poel, C.; Murphy, R.M. Benefits of prenatal taurine supplementation in preventing the onset of acute damage in the Mdx mouse. *PLoS Curr.* **2017**, *9*, 29188135.
79. Horvath, D.M.; Murphy, R.M.; Mollica, J.P.; Hayes, A.; Goodman, C.A. The effect of taurine and β-alanine supplementation on taurine transporter protein and fatigue resistance in skeletal muscle from mdx mice. *Amino Acids* **2016**, *48*, 2635–2645. [CrossRef]
80. Terrill, J.R.; Grounds, M.D.; Arthur, P.G. Increased taurine in pre-weaned juvenile mdx mice greatly reduces the acute onset of myofibre necrosis and dystropathology and prevents inflammation. *PLoS Curr.* **2016**, *8*, 27679740. [CrossRef] [PubMed]
81. Terrill, J.R.; Pinniger, G.J.; Graves, J.A.; Grounds, M.D.; Arthur, P.G. Increasing taurine intake and taurine synthesis improves skeletal muscle function in the mdx mouse model for Duchenne Muscular Dystrophy. *J. Physiol.* **2016**, *594*, 3095–3110. [CrossRef] [PubMed]
82. Bank, W.J.; Rowland, L.P.; Ipsen, J. Amino acids of plasma and urine in diseases of muscle. *Arch. Neurol.* **1971**, *24*, 176–186. [CrossRef] [PubMed]
83. De Paepe, B.; Martin, J.J.; Herbelet, S.; Jimenez-Mallebrera, C.; Iglesias, E.; Jou, C.; De Bleecker, J.L. Activation of osmolyte pathways in inflammatory myopathy and Duchenne Muscular Dystrophy points to osmoregulation as a contributing pathogenic mechanism. *Lab. Investig.* **2016**, *96*, 872–884. [CrossRef]
84. Kelly-Worden, M.; Thomas, E. Mitochondrial dysfunction in Duchenne Muscular Dystrophy. *J. Endocr. Metab. Dis.* **2014**, *4*, 211–218. [CrossRef]
85. Onopiuk, M.; Brutkowski, W.; Wierzbicka, K.; Wojciechowska, S.; Szczepanowska, J.; Fronk, J.; Zabłocki, K. Mutation in dystrophin-encoding gene affects energy metabolism in mouse myoblasts. *Biochem. Biophys. Res. Commun.* **2009**, *386*, 463–466. [CrossRef]
86. Jong, C.J.; Ito, T.; Mozaffari, M.; Azuma, J.; Schaffer, S. Effect of β-alanine treatment on mitochondrial taurine level and 5-taurinomethyluridine content. *J. Biomed. Sci.* **2010**, *17*, S25. [CrossRef]
87. McCarter, G.C.; Steinhardt, R.A. Increased activity of calcium leak channels caused by proteolysis near sarcolemmal ruptures. *J. Membr. Biol.* **2000**, *176*, 169–174. [CrossRef]
88. Matsumura, C.Y.; Taniguti, A.P.T.; Pertille, A.; Neto, H.S.; Marques, M.J. Stretch-activated calcium channel protein TRPC1 is correlated with the different degrees of the dystrophic phenotype in mdx mice. *Am. J. Physiol. Cell Physiol.* **2011**, *301*, C1344–C1350. [CrossRef] [PubMed]

89. Bellinger, A.M.; Reiken, S.; Carlson, C.; Mongillo, M.; Liu, X.; Rothman, L.; Marks, A.R. Hypernitrosylated ryanodine receptor calcium release channels are leaky in dystrophic muscle. *Nat. Med.* **2009**, *15*, 325–330. [CrossRef]
90. Mareedu, S.; Million, E.D.; Duan, D.; Babu, G.J. Abnormal calcium handling in Duchenne Muscular Dystrophy: Mechanisms and potential therapies. *Front. Physiol.* **2021**, *12*, 33897454. [CrossRef]
91. Wrogemann, K.; Pena, S.D.J. Mitochondrial calcium overload: A general mechanism for cell-necrosis in muscle diseases. *Lancet* **1976**, *307*, 672–674. [CrossRef]
92. Kyrychenko, V.; Poláková, E.; Janíček, R.; Shirokova, N. Mitochondrial dysfunctions during progression of dystrophic cardiomyopathy. *Cell Calcium* **2015**, *58*, 186–195. [CrossRef] [PubMed]
93. Peng, T.I.; Jou, M.J. Oxidative stress caused by mitochondrial calcium overload. *Ann. N. Y. Acad. Sci.* **2010**, *1201*, 183–188. [CrossRef] [PubMed]
94. Dubinin, M.V.; Talanov, E.Y.; Tenkov, K.S.; Starinets, V.S.; Mikheeva, I.B.; Sharapov, M.G.; Belosludtsev, K.N. Duchenne Muscular Dystrophy is associated with the inhibition of calcium uniport in mitochondria and an increased sensitivity of the organelles to the calcium-induced permeability transition. *Biochim. Biophys. Acta Mol. Basis Dis.* **2020**, *1866*, 165674. [CrossRef]
95. Ommati, M.M.; Farshad, O.; Jamshidzadeh, A.; Heidari, R. Taurine enhances skeletal muscle mitochondrial function in a rat model of resistance training. *PharmaNutrition* **2019**, *9*, 100161. [CrossRef]
96. Capogrosso, R.F.; Cozzoli, A.; Mantuano, P.; Camerino, G.M.; Massari, A.M.; Sblendorio, V.T.; De Luca, A. Assessment of resveratrol, apocynin and taurine on mechanical-metabolic uncoupling and oxidative stress in a mouse model of Duchenne Muscular Dystrophy: A comparison with the gold standard, α-methyl prednisolone. *Pharmacol. Res.* **2016**, *106*, 101–113. [CrossRef]
97. Barker, R.G.; Van der Poel, C.; Horvath, D.; Murphy, R.M. Taurine and methylprednisolone administration at close proximity to the onset of muscle degeneration is ineffective at attenuating force loss in the hind-limb of 28 days mdx mice. *Sports* **2018**, *6*, 109. [CrossRef]
98. Terrill, J.R.; Pinniger, G.J.; Nair, K.V.; Grounds, M.D.; Arthur, P.G. Beneficial effects of high dose taurine treatment in juvenile dystrophic mdx mice are offset by growth restriction. *PLoS ONE* **2017**, *12*, e0187317. [CrossRef] [PubMed]
99. De Luca, A.; Pierno, S.; Liantonio, A.; Cetrone, M.; Camerino, C.; Fraysse, B.; Camerino, D.C. Enhanced dystrophic progression in mdx mice by exercise and beneficial effects of taurine and insulin-like growth factor-1. *J. Pharmacol. Exp. Ther.* **2003**, *304*, 453–463. [CrossRef] [PubMed]
100. Giustarini, D.; Dalle-Donne, I.; Milzani, A.; Rossi, R. Low molecular mass thiols, disulfides and protein mixed disulfides in rat tissues: Influence of sample manipulation, oxidative stress and ageing. *Mech. Ageing Dev.* **2011**, *132*, 141–148. [CrossRef] [PubMed]
101. De Luca, A.; Pierno, S.; Liantonio, A.; Cetrone, M.; Camerino, C.; Simonetti, S.; Camerino, D.C. Alteration of excitation-contraction coupling mechanism in extensor digitorum longus muscle fibres of dystrophic mdx mouse and potential efficacy of taurine. *Br. J. Pharmacol.* **2001**, *132*, 1047–1054. [CrossRef]
102. Goodman, C.A.; Horvath, D.; Stathis, C.; Mori, T.; Croft, K.; Murphy, R.M.; Hayes, A. Taurine supplementation increases skeletal muscle force production and protects muscle function during and after high-frequency in vitro stimulation. *J. Appl. Physiol.* **2009**, *107*, 144–154. [CrossRef] [PubMed]
103. Cozzoli, A.; Rolland, J.F.; Capogrosso, R.F.; Sblendorio, V.T.; Longo, V.; Simonetti, S.; De Luca, A. Evaluation of potential synergistic action of a combined treatment with alpha-methyl-prednisolone and taurine on the mdx mouse model of Duchenne Muscular Dystrophy. *Neuropathol. Appl. Neurobiol.* **2011**, *37*, 243–256. [CrossRef] [PubMed]
104. Hirata, H.; Ueda, S.; Ichiseki, T.; Shimasaki, M.; Ueda, Y.; Kaneuji, A.; Kawahara, N. Taurine inhibits glucocorticoid-induced bone mitochondrial injury, preventing osteonecrosis in rabbits and cultured osteocytes. *Int. J. Mol. Sci.* **2020**, *21*, 6892. [CrossRef]
105. Hanaa, H.; Hamza, A.H. Potential role of arginine, glutamine and taurine in ameliorating osteoporotic biomarkers in ovariectomized rats. *Rep. Opin* **2009**, *1*, 24–35.
106. Campos, L.; Miziara, L.N.B.; Gallottini, M.; Ortega, K.; Martins, F. Medication-related osteonecrosis of the jaw in a Duchenne Muscular Dystrophy patient. *Photodiagnosis Photodyn. Ther.* **2020**, *31*, 101826. [CrossRef] [PubMed]
107. Dawson, R., Jr.; Biasetti, M.; Messina, S.; Dominy, J. The cytoprotective role of taurine in exercise-induced muscle injury. *Amino Acids* **2002**, *22*, 309–324. [CrossRef] [PubMed]
108. Miyazaki, T.; Matsuzaki, Y.; Ikegami, T.; Miyakawa, S.; Doy, M.; Tanaka, N.; Bouscarel, B. Optimal and effective oral dose of taurine to prolong exercise performance in rat. *Amino Acids* **2004**, *27*, 291–298. [CrossRef] [PubMed]
109. Yatabe, Y.; Miyakawa, S.; Miyazaki, T.; Matsuzaki, Y.; Ochiai, N. Effects of taurine administration in rat skeletal muscles on exercise. *J. Orthop. Sci.* **2003**, *8*, 415–419. [CrossRef]
110. Ma, Y.; Maruta, H.; Sun, B.; Wang, C.; Isono, C.; Yamashita, H. Effects of long-term taurine supplementation on age-related changes in skeletal muscle function of Sprague–Dawley rats. *Amino Acids* **2021**, *53*, 159–170. [CrossRef]
111. Galloway, S.D.; Talanian, J.L.; Shoveller, A.K.; Heigenhauser, G.J.; Spriet, L.L. Seven days of oral taurine supplementation does not increase muscle taurine content or alter substrate metabolism during prolonged exercise in humans. *J. Appl. Physiol.* **2008**, *105*, 643–651. [PubMed]
112. Shao, A.; Hathcock, J.N. Risk assessment for the amino acids taurine, L-glutamine and L-arginine. *Regul. Toxicol. Pharmacol.* **2008**, *50*, 376–399. [CrossRef]

113. Jeejeebhoy, F.; Keith, M.; Freeman, M.; Barr, A.; McCall, M.; Kurian, R.; Errett, L. Nutritional supplementation with MyoVive repletes essential cardiac myocyte nutrients and reduces left ventricular size in patients with left ventricular dysfunction. *Am. Heart J.* **2002**, *143*, 1092–1100. [CrossRef] [PubMed]
114. Suliman, M.E.; Bárány, P.; Filho, J.C.D.; Lindholm, B.; Bergström, J. Accumulation of taurine in patients with renal failure. *Nephrol. Dial. Transplant.* **2002**, *17*, 528–529. [CrossRef]
115. Brumbaugh, D.; Watne, L.; Gottrand, F.; Gulyas, A.; Kaul, A.; Larson, J.; Tomezsko, J. Nutritional and gastrointestinal management of the patient with Duchenne Muscular Dystrophy. *Pediatrics* **2018**, *142* (Suppl. 2), S53–S61. [CrossRef] [PubMed]
116. Davis, J.; Samuels, E.; Mullins, L. Nutrition considerations in Duchenne Muscular Dystrophy. *Nutr. Clin. Pract.* **2015**, *30*, 511–521. [CrossRef] [PubMed]
117. Liu, Y.; Mao, X.; Yu, B.; He, J.; Zheng, P.; Yu, J.; Chen, D. Excessive dietary taurine supplementation reduces growth performance, liver and intestinal health of weaned pigs. *Livest. Sci.* **2014**, *168*, 109–119. [CrossRef]
118. Boccanegra, B.; Verhaart, I.E.; Cappellari, O.; Vroom, E.; De Luca, A. Safety issues and harmful pharmacological interactions of nutritional supplements in Duchenne Muscular Dystrophy: Considerations for standard of care and emerging virus outbreaks. *Pharmacol. Res.* **2020**, *158*, 104917. [CrossRef]
119. Birnkrant, D.J.; Bushby, K.; Bann, C.M.; Apkon, S.D.; Blackwell, A.; Brumbaugh, D. DMD care considerations working group. Diagnosis and management of Duchenne Muscular Dystrophy, part 1: Diagnosis, and neuromuscular, rehabilitation, endocrine, and gastrointestinal and nutritional management. *Lancet Neurol.* **2018**, *17*, 251–267. [CrossRef]
120. Spurney, C.F.; Rocha, C.T.; Henricson, E.; Florence, J.; Mayhew, J.; Gorni, K. Cooperative international neuromuscular research group (CINRG) investigators. CINRG pilot trial of coenzyme Q10 in steroid-treated Duchenne Muscular Dystrophy. *Muscle Nerve* **2011**, *44*, 174–178. [CrossRef]

Article

N-acetyltaurine and Acetylcarnitine Production for the Mitochondrial Acetyl-CoA Regulation in Skeletal Muscles during Endurance Exercises

Teruo Miyazaki [1,*], Yuho Nakamura-Shinya [2], Kei Ebina [3], Shoichi Komine [4,5], Song-Gyu Ra [6], Keisuke Ishikura [7], Hajime Ohmori [8] and Akira Honda [1,9]

1 Joint Research Center, Tokyo Medical University Ibaraki Medical Center, Ami, Tokyo 300-0395, Japan; akihonda@tokyo-med.ac.jp
2 Department of Immunology, Faculty of Medicine, University of Tsukuba, Tsukuba 305-8575, Japan; yuhojump@yahoo.co.jp
3 Department of Health and Nutrition, Tsukuba International University, Tsuchiura 300-0051, Japan; ebinakei@gmail.com
4 Department of Acupuncture and Moxibustion, Faculty of Human Care, Teikyo Heisei University, Toshima-ku, Tokyo 170-8445, Japan; s.komine@thu.ac.jp
5 Faculty of Medicine, University of Tsukuba, Tsukuba 305-0821, Japan
6 Institute of Liberal Arts and Sciences, Tokushima University, Tokushima 770-8502, Japan; songgyura@tokushima-u.ac.jp
7 Faculty of Management, Josai University, Sakado 350-0295, Japan; ishikura@josai.ac.jp
8 Faculty of Health and Sport Sciences, University of Tsukuba, Tsukuba 305-8574, Japan; omori.hajime.gb@u.tsukuba.ac.jp
9 Department of Gastroenterology and Hepatology, Tokyo Medical University Ibaraki Medical Center, Ami, Tokyo 300-0395, Japan
* Correspondence: teruom@tokyo-med.ac.jp

Citation: Miyazaki, T.; Nakamura-Shinya, Y.; Ebina, K.; Komine, S.; Ra, S.-G.; Ishikura, K.; Ohmori, H.; Honda, A. N-acetyltaurine and Acetylcarnitine Production for the Mitochondrial Acetyl-CoA Regulation in Skeletal Muscles during Endurance Exercises. *Metabolites* **2021**, *11*, 522. https://doi.org/10.3390/metabo11080522

Academic Editor: Cholsoon Jang

Received: 16 July 2021
Accepted: 2 August 2021
Published: 6 August 2021

Abstract: During endurance exercises, a large amount of mitochondrial acetyl-CoA is produced in skeletal muscles from lipids, and the excess acetyl-CoA suppresses the metabolic flux from glycolysis to the TCA cycle. This study evaluated the hypothesis that taurine and carnitine act as a buffer of the acetyl moiety of mitochondrial acetyl-CoA derived from the short- and long-chain fatty acids of skeletal muscles during endurance exercises. In human subjects, the serum concentrations of acetylated forms of taurine (NAT) and carnitine (ACT), which are the metabolites of acetyl-CoA buffering, significantly increased after a full marathon. In the culture medium of primary human skeletal muscle cells, NAT and ACT concentrations significantly increased when they were cultured with taurine and acetate or with carnitine and palmitic acid, respectively. The increase in the mitochondrial acetyl-CoA/free CoA ratio induced by acetate and palmitic acid was suppressed by taurine and carnitine, respectively. Elevations of NAT and ACT in the blood of humans during endurance exercises might serve the buffering of the acetyl-moiety in mitochondria by taurine and carnitine, respectively. The results suggest that blood levels of NAT and ACT indicate energy production status from fatty acids in the skeletal muscles of humans undergoing endurance exercise.

Keywords: taurine; NAT; carnitine; ACT; acetyl-CoA; mitochondria; endurance exercise; skeletal muscle

1. Introduction

During energy production, mitochondrial acetyl-CoA is the metabolic junction that connects the TCA cycle with the upstream metabolic pathways of nutrient energy fuels, including carbohydrates, lipids, and amino acids. During endurance exercises, large amounts of ATP are produced from the acetyl-CoA in the mitochondria of skeletal muscles to meet the increasing energy demand [1,2]. An elevation in the mitochondrial acetyl-CoA/free-CoA ratio, accompanied by an increase in acetyl-CoA production, would suppress the metabolic flux from glycolysis to the TCA cycle in the skeletal muscles, because acetyl-CoA is an allosteric inhibitor of pyruvate dehydrogenase (PDH) (Figure 1) [3]. This feedback system is the role metabolic switches play to efficiently produce energy depending on the availability of energy fuels.

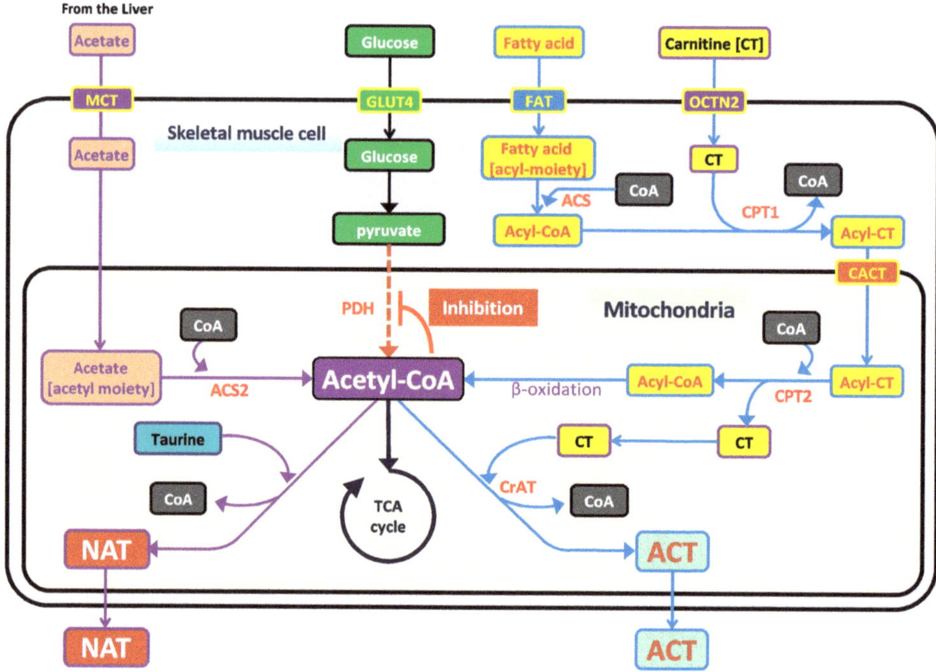

Figure 1. The preventive mechanisms for acetyl-CoA accumulation in the mitochondria through the acetylation of taurine and carnitine. An excess accumulation of mitochondrial acetyl-CoA metabolized from fatty acid and acetate would cause the metabolic stagnation of the glycolytic pathway as negative feedback through allosteric inhibition of PDH. To prevent the accumulation of mitochondrial acetyl-CoA, taurine and carnitine play buffering roles on the acetyl moiety of mitochondrial acetyl-CoA. Consequently, ACT and NAT are produced from fatty acids and acetate, respectively, and then excreted to extracellular fluids. Abbreviations: ACT; acetylcarnitine, ACS; acyl-CoA synthetase, ACS2; acetyl-CoA synthetase 2, CACT; carnitine acylcarnitine translocase, CPT; carnitine palmitoyl transferase, CrAT; carnitine acetyltransferase, CT; carnitine, GLUT4; glucose transporter 4, MCT; monocarboxylate transporter, NAT; *N*-acetyltaurine, OCTN2; organic anion transporter 2, PDH; pyruvate dehydrogenase.

In long-chain fatty acid β-oxidation, carnitine (β-hydroxy-γ-*N*-trimethylaminobutyric acid) plays important roles as an essential carrier for the transport of long-chain fatty acids from the cytoplasm into the mitochondrial matrix (Figure 1) and a physiological modulator of the mitochondrial pool of acetyl-CoA; i.e., the acetyl moiety of acetyl-CoA is converted to carnitine by carnitine acetyltransferase (CAT), which is located within the mitochondrial matrix, and, consequently, yields acetylcarnitine (ACT) and free-CoA [4]. Thereafter, ACT is excreted into the extracellular fluid. As the β-oxidation is activated for large energy production in endurance exercises, we suggest that the level of ACT in the blood increases during endurance exercises.

In addition, acetyl-CoA is produced in the mitochondria of skeletal muscles from the liver-supplied lipid metabolites, such as ketone bodies and a short-chain fatty acid, acetate. Although the utilization of ketone bodies in the skeletal muscles for energy production in endurance exercises has been well-known, acetate is also utilized as an efficient substrate for the production of mitochondrial acetyl-CoA via one metabolic reaction by acetyl-CoA synthetase 2 (ACS2) (Figure 1) [5–8]. Therefore, the acetyl-CoA derived from acetate is suggested to accumulate in the mitochondria during endurance exercises.

In a study by Shi et al. [9], the acetate yielded from ethanol detoxication in the liver reacted with taurine (2-aminoethanesulfonic acid), consequently forming *N*-acetyltaurine (NAT) for the facilitation of acetate excretion in the urine. Taurine exists in high concentrations in the whole body, particularly in the skeletal muscles and the liver [10–12], and it exhibits many physiological functions [13–17]. Taurine has been reported to enhance exercise performance in humans [18] and rodents [19–21], and its concentration in the skeletal muscles of rats significantly and time-dependently decreased during endurance exercises [13–17]. However, the reasons why taurine concentration in the skeletal muscle was significantly decreased after the endurance exercise and how taurine contributed to the exercise performance have been still unclear. We suggested that taurine in the skeletal muscles is utilized for NAT production and the NAT is excreted into the blood circulation during endurance exercises for the prevention of the accumulation of acetyl-CoA derived from acetate in the mitochondria.

Thus, we hypothesized that both taurine and carnitine may play roles in the regulations of the muscular mitochondrial amount of acetyl-CoA derived from short- and long-chain fatty acids in endurance exercises. The present study proposed the evaluation of the blood levels of NAT and ACT in humans before and after endurance exercises, and, furthermore, the investigation of the regulatory roles of taurine and carnitine in the mitochondrial balance of acetyl-CoA and free-CoA via the production of NAT and ACT using cultured skeletal muscle cells derived from humans.

2. Results

2.1. Serum Concentrations of Taurine, Carnitine, Its Acetylated Forms, and Energy Substrates during Endurace Exercises

Serum concentrations of taurine, NAT, carnitine, ACT, and other energy substrates were evaluated one day before (Pre), immediately after (Post), and the day after (Next) a full-marathon race. The taurine concentration was slightly higher in Post, but there was no significant difference among the three points (Figure 2A). On the contrary, the NAT concentration was significantly increased in Post compared to that in Pre, and the concentration in Next decreased significantly to the level in Pre (Figure 2B).

Similar to taurine concentration, there was no significant change in serum carnitine concentration before and after the exercise (Figure 2C). Serum ACT concentration was markedly and significantly higher in Post than in Pre, and it in Next, significantly decreased to the level in Pre (Figure 2D).

Serum concentrations of lactate, free fatty acid (FFA), and β-hydroxybutyrate significantly increased in Post compared to those in Pre (Table 1). On the next day, these energy-related parameters recovered to the level in Pre. There were no significant differences in serum glucose concentration among the three points (Table 1).

Table 1. Serum levels of biochemical energy-related parameters before and after the exercises.

Parameter	Pre	Post	Next
Glucose (mg/dL)	86.5 ± 1.0	83.0 ± 3.0	81.0 ± 1.7
Lactate (mg/dL)	18.1 ± 1.5	43.3 ± 3.9 *aa*	20.2 ± 1.0 *bb*
FFA (mEq/mL)	0.59 ± 0.08	1.73 ± 0.12 *aa*	0.64 ± 0.07 *bb*
β-hydroxybutyrate (μM)	48.2 ± 13.8	305.6 ± 40.1 *aa*	74.8 ± 15.6 *b*

Footnote: The data shown refer to the mean ± SEM. *aa*; $p < 0.0001$ compared to the Pre, *b*, and *bb*; $p < 0.001$ and $p < 0.0001$ compared to the Post by one-way ANOVA post-hoc multiple-comparison test. Abbreviations: FFA; free fatty acid, Pre; before the full marathon, Post; immediately after the full marathon, Next; the day after the full marathon.

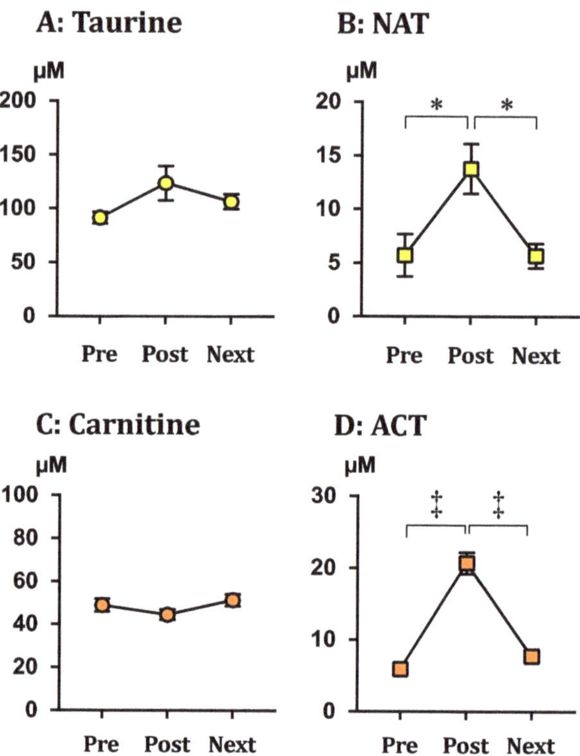

Figure 2. Serum levels of taurine (**A**), carnitine (**C**), and its acetylated forms, NAT (**B**) and ACT (**D**), before and after the endurance exercise. Serum was collected 1 day before (Pre), immediately after (Post), and the day after (Next) the full-marathon race. Taurine, NAT, carnitine, and ACT were simultaneously measured using an HPLC-MS/MS system. The data shown refer to the mean ± SEM. * $p < 0.05$ and ‡ $p < 0.0001$ were analyzed by one-way ANOVA post-hoc tests. Abbreviations: NAT; *N*-acetyltaurine, ACT; acetylcarnitine.

2.2. Intracellular and Extracellular Taurine, NAT, Carnitine, and ACT Concentrations in the Myotube Exposed to Acetate Combined with Taurine or Carnitine

To investigate whether the NAT and ACT are generated in the skeletal muscle cells by reaction with acetate and excreted to extracellular fluid, the myotube differentiated from human primary myoblast was exposed to sodium acetate following taurine or carnitine treatments, and then the NAT and ACT levels in the culture medium as well as in the cells were measured.

The myotube was exposed to 1 mM sodium acetate for 24 h following the pretreatments with 2 mM taurine or 300 µM carnitine for 24 h. Taurine level in the cell and culture medium was significantly increased by taurine treatment regardless of the presence of acetate (Figure 3A,B), while there was no significant difference for the exposure to acetate in the pretreatments of taurine and carnitine (Figure 3A,B). In both cell and culture medium, NAT concentration significantly increased under the taurine pretreatment conditions regardless of the presence of acetate compared to that under the respective untreated control condition (Figure 3C,D); however, under the taurine pretreatment condition, the concentration was significantly higher with acetate exposure than without (Figure 3C,D).

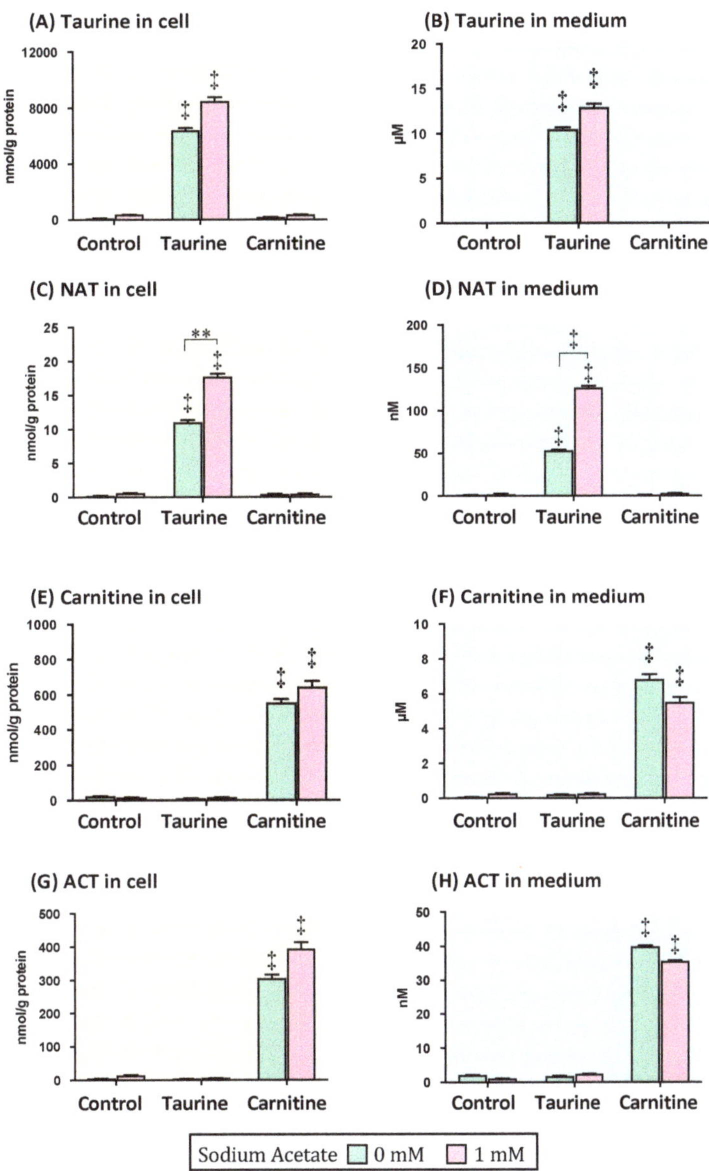

Figure 3. Taurine, carnitine, and its acetylated forms (NAT and ACT) in the myotube cell and culture medium after the treatments of taurine, carnitine, and acetate. The myotubes were treated with either 20 mM taurine or 300 μM carnitine for 24 h, and then exposed to 1 mM sodium acetate for 24 h. Thereafter, taurine, NAT, carnitine, and ACT concentrations in the cell (**A**,**C**,**E**,**G**) and culture medium (**B**,**D**,**F**,**H**) were measured. The data shown refer to the mean ± SEM. ** $p < 0.01$ and ‡ $p < 0.0001$ show the significant differences analyzed by the two-way ANOVA multiple comparison test. The symbols in the columns without bars show the significant difference compared to the respective control without taurine or carnitine treatment. Abbreviations: NAT; *N*-acetyltaurine, ACT; acetylcarnitine.

Carnitine and ACT concentrations in the cell and culture medium were significantly higher for the carnitine pretreatment than for the untreated control and taurine pretreatment, regardless of the exposure to acetate (Figure 3E–H). There were no significant differences in the carnitine and ACT concentrations of both the cell and culture mediums regardless of the presence of acetate (Figure 3E–H).

2.3. Extracellular NAT and ACT Concentrations in the Myotube Exposed to Palmitic Acid Combined with Taurine or/and Carnitine

In the next experiment, the myotube was used to investigate whether the NAT and ACT were generated in the skeletal muscle cells in the β-oxidation reaction and excreted to the extracellular fluid. The myotube was exposed to fatty acid following taurine or carnitine treatments, and then NAT and ACT levels in the culture medium were measured.

In the culture medium, NAT and ACT concentrations were measured after the exposure to 200 μM palmitic acid for 24 h following the pretreatments of 2 mM taurine and/or 50 or 300 μM carnitine for 24 h. NAT concentration significantly increased under the taurine pretreatment conditions (taurine alone and the combination with carnitine) regardless of the exposure to palmitic acid compared to that in the respective untreated control (Figure 4A). Particularly, in the combination of taurine and carnitine, the NAT concentration was significantly higher with palmitic acid exposure than without. There was no difference in NAT concentration of the pretreatment with taurine alone regardless of the exposure to palmitic acid. Under the conditions without taurine pretreatment (control and carnitine pretreatments), NAT in the culture medium was almost nondetectable.

On the contrary, ACT concentration significantly increased via the treatments with carnitine alone and together with taurine under the exposure to palmitic acid compared to those under the respective pretreatment conditions without palmitic acid exposure, with the untreated control, and with taurine and palmitic acid (Figure 4B). Under the condition with no palmitic acid exposure, ACT concentration was also significantly increased in the carnitine pretreatments (carnitine alone and combined with taurine) compared to that in the respective nontreatment.

2.4. Mitochondrial Acetyl-CoA/Free-CoA Ratio in the Myotube Exposed to Acetate or Palmitic Acid Combined with Taurine or Carnitine

For evaluations of the effects of taurine and carnitine on the mitochondrial acetyl-CoA/free-CoA ratio, the myotube was treated with 2 mM taurine or carnitine (50 or 300 μM) with the mitochondrial acetyl-CoA generative condition of 1 mM sodium acetate exposure or with the β-oxidation condition (200 μM palmitic acid with 50 μM carnitine).

Figure 5A shows the mitochondrial acetyl-CoA/free-CoA ratio under the sodium acetate exposure condition. The acetyl-CoA/free-CoA ratio significantly increased upon exposure to sodium acetate alone (Control) compared to that under the untreated condition without sodium acetate, taurine, or carnitine (NT). In the taurine treatment with acetate (Tau), the ratio did not increase, and it was maintained at the same level under the NT condition. On the contrary, the ratio significantly increased in the carnitine treatment with acetate (CT) compared to that under the NT and Tau conditions.

Figure 5B demonstrates the mitochondrial acetyl-CoA/free-CoA ratio in the fatty acid β-oxidation condition. The ratio significantly increased under the β-oxidation control condition (Control), which was the cotreatment with 200 μM palmitic acid and 50 μM carnitine, compared to those under the no treatment condition (NT), palmitic acid alone (PA), and combined with taurine (Tau). The result in the Control implies that the β-oxidation reaction progressed in the presence of carnitine and palmitic acid. There was no difference in the ratio among the carnitine untreated conditions (NT, PA, and Tau). On the contrary, the higher dose (300 μM) of carnitine with palmitic acid (CT) significantly suppressed the increase in ratio observed with the Control condition, while the taurine cotreatment with 50 μM carnitine (Combined) was lower, but not significantly, than in the Control.

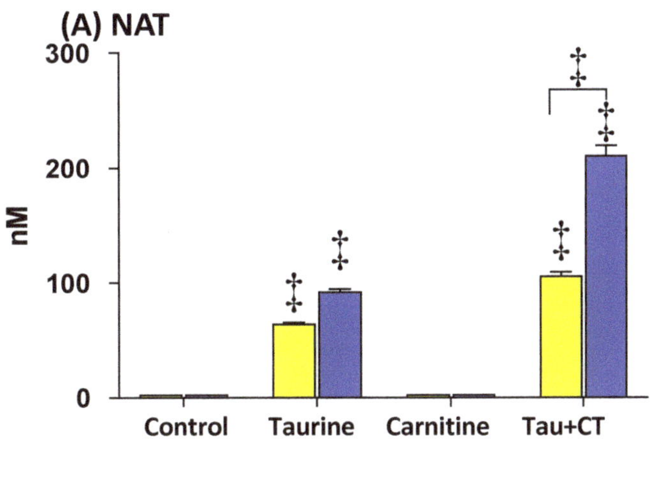

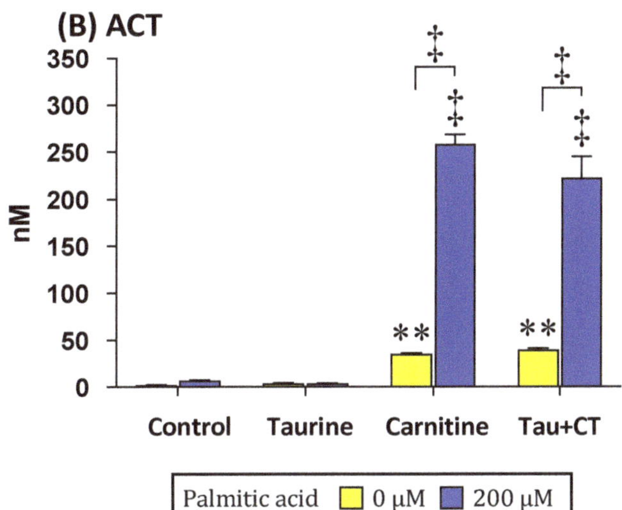

Figure 4. NAT and ACT concentrations in the culture medium of the myotube after exposure to palmitic acid combined with taurine and/or carnitine. The myotube was treated with 20 mM taurine or/and 300 μM carnitine for 24 h. Then, it was exposed to 200 μM palmitic acid for 24 h. NAT (**A**) and ACT (**B**) concentrations in the culture medium were measured. The data shown refer to the mean ± SEM. ** $p < 0.01$ and ‡ $p < 0.0001$ show the significant differences analyzed by the two-way ANOVA multiple comparison test. The symbols on the columns without bars show the significant difference compared to those in the non-pretreated control and taurine pretreated with palmitic acid. Abbreviations: NAT; *N*-acetyltaurine, ACT; acetylcarnitine, Tau+CT; taurine and carnitine.

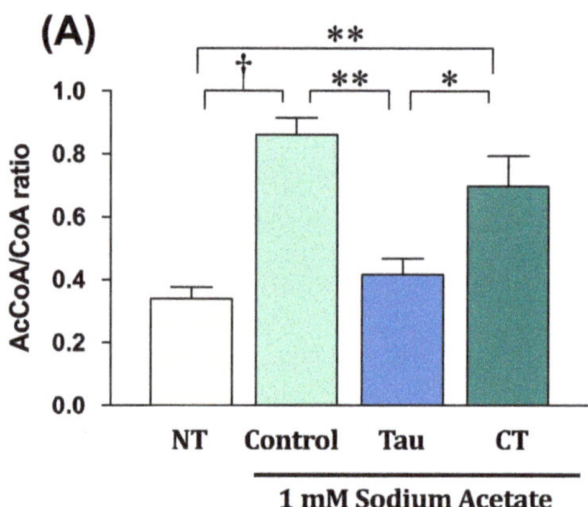

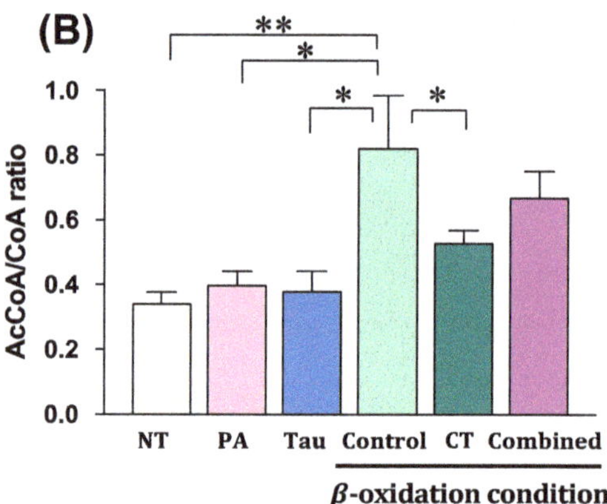

Figure 5. Mitochondrial acetyl-CoA/free-CoA ratio in the myotube exposed to acetate (**A**) and fatty acid (**B**) with taurine or/and carnitine treatments. (**A**) The myotube was exposed to 1 mM sodium acetate with 2 mM taurine or 300 μM carnitine for 24 h. Abbreviations: NT; no treatments with acetate, taurine, or carnitine, Control; 1 mM sodium acetate alone, Tau; 2 mM taurine with 1 mM sodium acetate, CT; 300 μM carnitine with 1 mM sodium acetate. (**B**) The myotube was exposed to 200 μM palmitic acid with 2 mM taurine, 50 μM carnitine, 300 μM carnitine, or a combination of 2 mM taurine and 50 μM carnitine for 24 h. Abbreviations: NT; no treatments with palmitic acid, taurine, or carntine, PA; 200 μM palmitic acid alone, Tau; 2 mM taurine with 200 μM palmitic acid, Control; 50 μM carnitine with 200 μM palmitic acid, CT; 300 μM carnitine with 200 μM palmitic acid, Combined; combination of 50 μM carnitine and 2 mM taurine with 200 μM palmitic acid, β-oxidation condition; culture condition with carnitine and palmitic acid. After these exposures for 24 h, acetyl-CoA and free CoA levels in the mitochondria were measured, and the ratio of acetyl-CoA to free CoA was calculated. The data shown refer to the mean ± SEM. * $p < 0.05$, ** $p < 0.01$, and † $p < 0.001$ show the significant differences analyzed by the one-way ANOVA post-hoc multiple comparison test.

3. Discussion

The present study evaluated the hypothesis that taurine and carnitine act as regulators of the mitochondrial acetyl-CoA/free-CoA ratio in the energy production from short- and long-chain fatty acids, via NAT and ACT formations in the skeletal muscles during endurance exercises. The results indicate that the serum levels of NAT and ACT significantly increased immediately after the full-marathon race. In the experiments of cultured skeletal muscle cells, the NAT was significantly excreted from the myotube to the culture medium when taurine was treated together with acetate, while significant ACT excretion was observed in the cotreatment of carnitine and palmitic acid. In addition, the significantly increased mitochondrial acetyl-CoA/free-CoA ratio seen for the treatments with acetate or fatty acid was significantly suppressed by cotreatments with taurine or carnitine, respectively.

Carnitine acts as the modulator of the accumulation of mitochondrial acetyl-CoA induced in fatty acid β-oxidation, excreting the consequent metabolite ACT from the skeletal muscles [1,2]. As fatty acid β-oxidation is activated to fulfill the high demand of energy production in the skeletal muscles during endurance exercises, we suggested in this study that the ACT concentration in blood elevates during the endurance exercises. In the results, serum concentrations of ACT, as well as β-hydroxybutyrate, significantly increased after the full-marathon race (Figure 2D and Table 1). The blood level of β-hydroxybutyrate, which is the most stable ketone body in blood, is established as a marker of hepatic fatty acid β-oxidation because the metabolite of fatty acid β-oxidation is excreted from the liver under ketogenic conditions such as during endurance exercises and starvation [22,23]. On the contrary, there are few studies in humans that illustrate the change in the blood ACT level during endurance exercises [24], while significant increases in the ACT levels in the skeletal muscles during exercise have been reported [1,25–27]. The formation of ACT in the skeletal muscles has been recognized as the marker for intramuscular ATP production via oxidative phosphorylation from glycolysis and fatty acid β-oxidation [26]. Based on the facts that most of the carnitine in the body is contained in the skeletal muscles [28] and carnitine is the essential regulator of the β-oxidation of skeletal muscles [29], the increased level of blood ACT is thought to reflect the increase in muscular β-oxidation activity in endurance exercises. This is also supported by the present results that show that ACT was excreted from the muscle cells exposed to the fatty acid together with carnitine (Figure 4B), along with the improvement in the mitochondrial acetyl-CoA/free-CoA ratio (Figure 5B).

Similar to ACT, a significantly increased NAT concentration in the serum was observed after the full-marathon race (Figure 2B). NAT has been reported to be a metabolite of taurine reacting with acetate, which is a product in the detoxification of ethanol in the liver, in order to facilitate the urinary excretion of excess acetate by increasing the hydrophilicities via reaction with taurine [9]. Acetate is also generated from acetyl-CoA in fatty acid β-oxidation, along with the ketone bodies in the liver under ketogenic conditions, and it is supplied as an energy fuel to extrahepatic tissues, including skeletal muscle. As acetate is metabolized directly back to the mitochondrial acetyl-CoA in one metabolic step by ACS2 (Figure 1) [5–8], acetate is thought to be a more efficient energy fuel than other nutrients, including glucose and fatty acids, which require multiple metabolic reactions to generate acetyl-CoA in skeletal muscles. However, it is possible for acetyl-CoA to accumulate in the mitochondria of skeletal muscle if acetate is excessively supplied from the liver during and after endurance exercises.

In addition to the abundance of taurine in the skeletal muscles [10–12] and the excretion of NAT from the skeletal muscle cells cultured with acetate and taurine, our previous study confirmed that plasma NAT concentration was increased, correlated with that in the skeletal muscle, but not in the liver, after a treadmill exercise in rats [30]. These findings strongly supported that the significantly increased NAT in serum was derived from the skeletal muscle during the endurance exercise in humans. Therefore, there is a possibility that the blood levels of NAT could be a potential biomarker of energy production status from short-chain fatty acids, similarly to ACT for long-chain fatty acids, during endurance exercise in the skeletal muscles of humans.

In the cell culture experiment, the acetyl-CoA/free CoA ratio in the mitochondria significantly increased upon acetate treatment (Figure 5A). Serum acetate concentration was likely to increase significantly along with the significant increase of β-hydroxybutyrate and

FFA after the endurance exercises (Figure 2). Therefore, it is suggested that the mitochondrial acetyl-CoA derived from acetate might have increased in the skeletal muscles after the full-marathon race. The present cell culture experiments also showed that the significantly increased acetyl-CoA/free-CoA ratio in the mitochondria of the myotube was suppressed by taurine treatment, and that the excretion of NAT to the culture medium significantly increased (Figure 3D). Indeed, the serum NAT concentration significantly increased after the full-marathon race, and we preliminarily reported the significant increase in urinary NAT excretion in a human after endurance exercise [31]. Our data support the hypothesis that taurine acts as a buffer of the acetyl moiety of excess mitochondrial acetyl-CoA derived from acetate in the skeletal muscles during endurance exercises.

In the myotube experiments, NAT concentrations in both cell and culture medium were significantly higher in the taurine and acetate cotreatment than in the treatment with taurine alone (Figure 3C,D). On the contrary, ACT concentrations in both cell and culture mediums were significantly increased via carnitine treatment (Figure 3G,H), but the excretion level was not influenced in the presence of acetate. Thus, these results imply that ACT is not generated from the acetyl-CoA that is derived from acetate. In addition, the NAT excretion to the culture medium significantly increased via taurine treatment under the β-oxidation condition (Figure 4A), implying that taurine contributes to the regulation of the mitochondrial acetyl-CoA/free-CoA ratio in the fatty acid β-oxidation.

Together with the many physiological actions [13–17], including the enhancement of exercise performance [18–20], taurine may have beneficial effects against physical fatigue and recovery by exercise. The present study shows that taurine plays a preventive role in the increase in the mitochondrial acetyl-CoA/free-CoA ratio derived from acetate in the skeletal muscles by the generation of NAT, similar to the role carnitine plays in β-oxidation, which results in ACT generation. The mitochondrial acetyl-CoA accumulation leads to the delay in the glycolysis pathway and insulin resistance in the skeletal muscles; the reduction in PDH activity by mitochondrial acetyl-CoA delays the metabolic flux from glycolysis to the TCA cycle [32], and the delayed flux reduces the phosphofructokinase activity and accumulates glucose-6-phosphate, which allosterically reduces the activity of hexokinase, which is the rate-limiting enzyme of glycolysis. This continuous negative feedback controlled by the cell energy state finally results in a decline in insulin-dependent glucose uptake [33–36]. In a previous study, we showed that insulin resistance was caused in the C2C12 myotube by excess ACT treatment that induces the acetyl-CoA accumulation in the mitochondria through the reverse transfer of the acetyl moiety to free-CoA [37]. Therefore, we speculated that the beneficial effects of taurine on fatigue and recovery after exercise is related to the improvement of the impaired glycolysis by the regulation of mitochondrial acetyl-CoA accumulation through NAT generation.

4. Materials and Methods

4.1. Materials

Taurine, L-carnitine, sodium acetate, acetic anhydride, sucrose, EDTA, potassium dihydrogen phosphate, acetonitrile, 2-propanol, methanol, 4-dimethylaminopyridine, and pyridine were purchased from FUJIFILM Wako Pure Chemical Corporation (Osaka, Japan). NAT was purchased from Carbosynth Limited (Berkshire, UK). Acetyl-L-carnitine HCl, dibutylammonium acetate (DBAA), 2-pyridinemethanol (2PM), and 2-methyl-6-nitrobenzoic anhydride were purchased from Tokyo Kasei Kogyo (Tokyo, Japan). [$^2H_4$]taurine (taurine-d4), acetyl-L-[$^2H_3$]carnitine HCl (ACT-d3), L-[$^2H_3$]carnitine HCl (carnitine-d3), and DL-lactate-[$^2H_3$] (lactate-d3) were obtained from C/D/N Isotopes Inc. (Quebec, Canada). NAT-d4 was synthesized from taurine-d4 by the reaction with acetic anhydride [38]. Taurine-d4, NAT-d4, carnitine-d3, and ACT-d3 were mixed with acetonitrile to use as an internal standard (IS) mixture. Acetyl-CoA, HMG-CoA, free-CoA, sodium [$^{13}C_4$]DL-β-hydroxybutyrate (βHB-$^{13}C_4$), and bovine serum albumin (BSA) were obtained from Merck, KGaA (Darmstadt, Germany). Human primary skeletal muscle myoblasts isolated from the rectus abdominus muscle, and growth medium (GM) and differentiation medium (DM) specialized for human cells were purchased from ZenBio, Inc. (Durham, UK). All of the other reagents for the cell culture experiments were purchased from Thermo Fisher Scientific (Gibco®, Waltham, MA, USA). The solvents used for analysis were of analytical grade.

4.2. Exercise Experiment in Humans

In the human study, 17 healthy male volunteers who participated in the 32nd Tsukuba full marathon (held in Ibaraki, Japan) were recruited. The average goal time of the subjects was 264.6 ± 60.6 min (mean ± standard error [SEM]). In the subjects, blood was collected at three points: One day before (Pre), immediately after (Post), and the day after (Next) the full-marathon race. The serum was maintained at −20 °C until analysis.

4.3. Differentiation of Cultured Skeletal Muscle Cell and Treatments

For the experiments of skeletal muscle cell culture, human primary skeletal muscle myoblasts were used followed by differentiation to myotubes according to the manufacture by ZenBio, Inc. The myoblasts seeded in a 12-well plate were cultured with the GM until confluent and were then differentiated to the myotube with the DM for 5 days. The cells were maintained at 37 °C in a humid atmosphere of 5% CO_2.

For the acetate exposure experiment, the myotube was exposed to 20 mM taurine or 300 µM carnitine in the DM for 24 h. Doses of taurine and carnitine were referenced with our previous studies [37,39]. After washing with PBS twice, the myotube was further exposed to 1 mM sodium acetate in GM for an additional 24 h.

For the fatty acid exposure experiment, the myotube was exposed to 20 mM taurine and/or 300 µM carnitine in the DM for 24 h; then, it was exposed to 200 µM palmitic acid dissolved in a solution of 10% BSA-PBS in the DMEM including 1 g/L glucose without FBS for an additional 24 h.

After both experiments, the myotube and culture medium were collected for analyses of taurine, NAT, carnitine, ACT, free-CoA, and acetyl-CoA using an HPLC-MS/MS system. Data were obtained from four independent experiments (n = 4).

4.4. Taurine, NAT, Carnitine, and ACT Analyses

Taurine, NAT, carnitine, and ACT in samples were quantified by an HPLC-MS/MS system according to our previous study [37]. The myotube cells collected in a collection tube were sonicated in 500 µL of KP buffer (10 mM potassium-phosphate buffer, pH 5.5). After centrifugation at 3500× g at 4 °C for 10 min, the supernatant was used for analysis. The protein concentration in the supernatant was measured using a Pierce® BCA Protein Assay Kit (Thermo Fischer Scientific Inc., Rockford, MI, USA). Fifty microliters of the culture medium and the supernatants from the cell, and 5 µL of human serum were mixed with 50 µL of IS solution containing 100 ng of taurine-d4, 1 ng of NAT-d4, 5 ng of carnitine-d3, and 5 ng of ACT-d3 in acetonitrile–water (19:1, v/v). After centrifugation at 2000× g for 1 min, the supernatant was evaporated to dryness at 80 °C under a nitrogen stream. The residue was redissolved in 60 µL of 0.1% formic acid, and an aliquot (5 µL) was injected into the LC-MS/MS system and was analyzed in electrospray ionization (ESI) mode.

Chromatographic separation was performed using a Hypersil GOLD aQ column (150 × 2.1 mm, 3 µm, Thermo Fischer Scientific Inc.) at 40 °C. The mobile phase was methanol–water (1:9, v/v) containing 0.1% formic acid, at a flow rate of 200 µL/min. The MS/MS conditions were spray voltage, 3000 V; vaporizer temperature, 450 °C; sheath gas (nitrogen) pressure, 50 psi; auxiliary gas (nitrogen) flow, 15 arbitrary units; ion transfer capillary temperature, 220 °C; collision gas (argon) pressure, 1.0 motor; collision energy, 20 V; ion polarity, positive; selected reaction monitoring (SRM), m/z 124 → m/z 80 for the taurine, m/z 128 → m/z 80 for the taurine-d4 variant, m/z 166 → m/z 80 for the NAT, m/z 170 → m/z 80 for the NAT-d4 variant, m/z 162 → m/z 103 for the carnitine, m/z 165 → m/z 103 for the carnitine-d3 variant, m/z 204 → m/z 85 for the ACT, and m/z 207 →

m/*z* 85 for the ACT-d3 variant. The analyzed results for the cell sample are expressed per protein.

4.5. Other Blood Biochemical Analyses

In serum, glucose and FFA concentrations were measured using commercially available kits: Glucose CII-test Wako and NEFA C-test kit Wako (FUJIFILM Wako Pure Chemical Corporation).

Lactate and β-hydroxybutyrate were analyzed using the HPLC-MS/MS system with derivatization into 2PM esters according to our previous studies [22,40]. Serum (5 μL) was mixed with 200 μL of acetonitrile–water (19:1, *v*/*v*) containing 100 ng of βHB-$^{13}C_4$ and 500 ng of lactate-d3 as an IS. The supernatant obtained after centrifugation at 2000× *g* for 1 min was evaporated to dryness at 80 °C under a nitrogen stream. The residue was resolved with a 50 μL mixture of reagents containing 2-methyl-6-nitrobenzoic anhydride, 4-dimethylaminopyridine, pyridine, and 2-PM, and it was incubated at room temperature for 30 min. Next, the reactant was centrifugated at 700× *g* for 1 min followed by mixing with 1 mL of diethyl ether, and the supernatant was evaporated at 55 °C under a nitrogen stream. The residue redissolved in 100 μL of 0.1% formic acid–water (*v*/*v*), and 5 μL of the supernatant after spin down was injected into the HPLC-MS/MS system. Chromatographic separation was performed using the Hypersil GOLD aQ column at 40 °C. The mobile phase was initially the acetonitrile–water containing 0.2% formic acid at a flow rate of 300 μL/min for 5 min; then, the flow rate was switched to 300 μL/min for an additional 7 min. The general MS/MS conditions were as follows: Spray voltage, 3000 V; vaporizer temperature, 450 °C; sheath gas (nitrogen) pressure, 50 psi; auxiliary gas (nitrogen) flow, 15 arbitrary units; ion transfer capillary temperature, 220 °C; collision gas (argon) pressure, 1.0 mTorr; collision energy, 15 V; ion polarity, positive; SRM, *m*/*z* 196 → *m*/*z* 110 for the 2PM-β-hydroxybutyrate, *m*/*z* 200 → *m*/*z* 110 for the 2PM-βHB-$^{13}C_4$, *m*/*z* 182.1 → *m*/*z* 110.1 for the 2PM-Lactate, and *m*/*z* 185.1 → *m*/*z* 110.1 for the 2PM-Lactate-d3.

4.6. Mitochondrial Free-CoA and Acetyl-CoA Concentrations in Myotubes

The mitochondrial fraction was prepared from the collected myotubes according to the previously reported method [41], and acetyl-CoA and free-CoA contents were measured by our method reported previously [37]. The myotube was homogenized with a loose-fitting Teflon pestle in 4 volumes of 3 mM Tris-HCl buffer (pH 7.4) containing 0.25 mM sucrose and 0.1 mM EDTA, and it was centrifuged at 700× *g* for 10 min. The supernatant was further centrifuged at 7000× *g* for 20 min. After the centrifugation, the mitochondrial pellet was homogenized in 100 μL of freshly prepared potassium dihydrogen phosphate (pH 4.9) (100 mM) and 100 μL of acetonitrile, 2-propanol, and methanol (3:1:1, *v*/*v*/*v*) with 50 pmol of the HMG-CoA as an IS; then, it was centrifuged at 16,000× *g* at 4 °C for 10 min. The supernatant was dried at 50 °C under a nitrogen stream and was then resuspended in 50 μL of 50% methanol–water solution. After centrifugation at 14,000× *g* for 10 min at 4 °C, 5 μL of the supernatant was injected into the LC-MS/MS system and was analyzed in ESI mode for quantification of the mitochondrial acetyl-CoA and free-CoA.

Chromatographic separation was performed using the Hypersil GOLD aQ column at 40 °C, with the following gradient system at a flow rate of 200 μL/min. Initially, the mobile phase comprised 5 mM aqueous DBAA-methanol (4:1, *v*/*v*), which was programmed to change to 5 mM aqueous DBAA-methanol (3:7, *v*/*v*) in a linear manner over 10 min. The MS/MS conditions were spray voltage, 2500 V; vaporizer temperature, 450 °C; sheath gas (nitrogen) pressure, 50 psi; auxiliary gas (nitrogen) flow, 15 arbitrary units; ion transfer capillary temperature, 220 °C; collision gas (argon) pressure, 1.0 motor; collision energy, 35 V; ion polarity, negative; SRM, *m*/*z* 766 → *m*/*z* 408, *m*/*z* 808 → *m*/*z* 408, and *m*/*z* 910.1 → *m*/*z* 408.1 for CoA, acetyl-CoA, and HMG-CoA, respectively. The ratio of acetyl-CoA to free-CoA was calculated from the value of mitochondrial acetyl-CoA and free-CoA contents.

4.7. Statistical Analysis

Statistical significance was determined via the unpaired Student's *t*-test, or one-way or two-way ANOVA multiple comparison tests. The data expressed refer to the mean ± SEM. Differences were considered statistically significant when the calculated *p* value was less than 0.05.

5. Conclusions

The present study confirmed that the serum NAT and ACT concentrations significantly increased after endurance exercises. In the cultured muscle cells, the increased mitochondrial acetyl-CoA/free-CoA ratios induced by acetate or palmitic acid exposures were significantly suppressed via taurine or carnitine treatment, respectively, together with the increase in extracellular NAT and ACT excretions. In conclusion, the present results support the hypothesis that taurine and carnitine act as regulators of the acetyl moiety of AcCoA derived from short- and long-chain fatty acids in the mitochondria of skeletal muscles during endurance exercises by excretions of NAT and ACT to blood from the skeletal muscle. In turn, the NAT and ACT production may support the regulation of energy metabolism during exercise and of recovery from muscle fatigue after endurance exercises. The present findings suggest that blood NAT and ACT levels could be the parameters of energy production status from acetate and fatty acids in the skeletal muscles in endurance exercise.

Author Contributions: Conceptualization, T.M. and A.H.; methodology, T.M., Y.N.-S., K.E., S.K., S.-G.R. and K.I.; validation, A.H.; formal analysis, T.M. and A.H.; data curation, T.M.; writing—original draft preparation, T.M.; writing—review and editing, A.H.; visualization, T.M.; supervision, A.H. and H.O.; project administration, T.M. and A.H.; funding acquisition, T.M. All authors have read and agreed to the published version of the manuscript.

Funding: This research was funded by KAKENHI grant (25750334) from the Japanese Society for the Promotion of Science.

Institutional Review Board Statement: The study was carried out in accordance with the Declaration of Helsinki and was approved by the Human Subjects Committee of the University of Tsukuba (Approved #TAI24-42). Informed consent was obtained from all subjects involved in the study.

Informed Consent Statement: Informed consent was obtained from all subjects involved in the study. Written informed consent has been obtained from the patient(s) to publish this paper.

Data Availability Statement: Not applicable.

Conflicts of Interest: The authors declare no competing interest.

Abbreviations

The following abbreviations are used in this manuscript:

βHB-^{13}C4	[$^{13}C_4$]β-hydroxybutyrate-^{13}C4
2PM	2-pyridinemethanol
ACT	acetylcarnitine
ACT-d3	acetyl-L-[$^2H_3$]carnitine HCl
ACS2	acetyl-CoA synthetase 2
BSA	bovine serum albumin
CACT	carnitine acylcarnitine translocase
carnitine-d3	L-[$^2H_3$]carnitine HCl
CAT	carnitine acetyltransferase
CPT	carnitine palmitoyl transferase
CrAT	carnitine acetyltransferase
CT	carnitine
DBAA	dibutylammonium acetate
DM	differentiation medium
DMEM	Dulbecco's modified Eagle's medium

ESI	electrospray ionization
FBS	Fetal bovine serum
FFA	free fatty acid
GLUT4	glucose transporter 4
GM	growth medium
IS	internal standard
lactate-d3	DL-lactate-[$^2H_3$]
MCT	monocarboxylate transporter
NAT	*N*-acetyltaurine
OCTN2	organic anion transporter 2
PDH	pyruvate dehydrogenase
SEM	standard error
taurine-d4	[$^2H_4$]taurine

References

1. Constantin-Teodosiu, D.; Carlin, J.I.; Cederblad, G.; Harris, R.C.; Hultman, E. Acetyl group accumulation and pyruvate dehydrogenase activity in human muscle during incremental exercise. *Acta Physiol. Scand.* **1991**, *143*, 367–372. [CrossRef] [PubMed]
2. Howlett, R.A.; Parolin, M.L.; Dyck, D.J.; Hultman, E.; Jones, N.L.; Heigenhauser, G.J.; Spriet, L.L. Regulation of skeletal muscle glycogen phosphorylase and PDH at varying exercise power outputs. *Am. J. Physiol.* **1998**, *275*, R418–R425. [CrossRef] [PubMed]
3. Randle, P.J.; Garland, P.B.; Hales, C.N.; Newsholme, E.A. The glucose fatty-acid cycle. Its role in insulin sensitivity and the metabolic disturbances of diabetes mellitus. *Lancet* **1963**, *1*, 785–789. [CrossRef]
4. Stephens, F.B.; Constantin-Teodosiu, D.; Greenhaff, P.L. New insights concerning the role of carnitine in the regulation of fuel metabolism in skeletal muscle. *J. Physiol.* **2007**, *581*, 431–444. [CrossRef] [PubMed]
5. Fukao, T.; Lopaschuk, G.D.; Mitchell, G.A. Pathways and control of ketone body metabolism: On the fringe of lipid biochemistry. *Prostaglandins Leukot. Essent. Fatty Acids* **2004**, *70*, 243–251. [CrossRef] [PubMed]
6. Luong, A.; Hannah, V.C.; Brown, M.S.; Goldstein, J.L. Molecular characterization of human acetyl-CoA synthetase, an enzyme regulated by sterol regulatory element-binding proteins. *J. Biol. Chem.* **2000**, *275*, 26458–26466. [CrossRef]
7. Fujino, T.; Kondo, J.; Ishikawa, M.; Morikawa, K.; Yamamoto, T.T. Acetyl-CoA synthetase 2, a mitochondrial matrix enzyme involved in the oxidation of acetate. *J. Biol. Chem.* **2001**, *276*, 11420–11426. [CrossRef]
8. Sakakibara, I.; Fujino, T.; Ishii, M.; Tanaka, T.; Shimosawa, T.; Miura, S.; Zhang, W.; Tokutake, Y.; Yamamoto, J.; Awano, M.; et al. Fasting-induced hypothermia and reduced energy production in mice lacking acetyl-CoA synthetase 2. *Cell Metab.* **2009**, *9*, 191–202. [CrossRef] [PubMed]
9. Shi, X.; Yao, D.; Chen, C. Identification of N-acetyltaurine as a novel metabolite of ethanol through metabolomics-guided biochemical analysis. *J. Biol. Chem.* **2012**, *287*, 6336–6349. [CrossRef]
10. Airaksinen, E.M.; Paljarvi, L.; Partanen, J.; Collan, Y.; Laakso, R.; Pentikainen, T. Taurine in normal and diseased human skeletal muscle. *Acta Neurol. Scand.* **1990**, *81*, 1–7. [CrossRef]
11. Awapara, J. The taurine concentration of organs from fed and fasted rats. *J. Biol. Chem.* **1956**, *218*, 571–576. [CrossRef]
12. Iwata, H.; Obara, T.; Kim, B.K.; Baba, A. Regulation of taurine transport in rat skeletal muscle. *J. Neurochem.* **1986**, *47*, 158–163. [CrossRef] [PubMed]
13. Pasantes, M.H.; Quesada, O.; Moran, J. Taurine: An osmolyte in mammalian tissues. *Adv. Exp. Med. Biol.* **1998**, *442*, 209–217. [CrossRef]
14. Huxtable, R.J. Physiological actions of taurine. *Physiol. Rev.* **1992**, *72*, 101–163. [CrossRef]
15. Miyazaki, T.; Matsuzaki, Y. Taurine and liver diseases: A focus on the heterogeneous protective properties of taurine. *Amino Acids* **2014**, *46*, 101–110. [CrossRef]
16. Nakamura, T.; Ogasawara, M.; Koyama, I.; Nemoto, M.; Yoshida, T. The protective effect of taurine on the biomembrane against damage produced by oxygen radicals. *Biol. Pharm. Bull.* **1993**, *16*, 970–972. [CrossRef] [PubMed]
17. Miyazaki, T. Involvement of taurine in the skeletal muscle on exercise. *Adv. Exerc. Sports Physiol.* **2010**, *15*, 131–134.
18. da Silva, L.A.; Tromm, C.B.; Bom, K.F.; Mariano, I.; Pozzi, B.; da Rosa, G.L.; Tuon, T.; da Luz, G.; Vuolo, F.; Petronilho, F.; et al. Effects of taurine supplementation following eccentric exercise in young adults. *Appl. Physiol. Nutr. Metab.* **2014**, *39*, 101–104. [CrossRef] [PubMed]
19. Miyazaki, T.; Matsuzaki, Y.; Ikegami, T.; Miyakawa, S.; Doy, M.; Tanaka, N.; Bouscarel, B. Optimal and effective oral dose of taurine to prolong exercise performance in rat. *Amino Acids* **2004**, *27*, 291–298. [CrossRef] [PubMed]
20. Yatabe, Y.; Miyakawa, S.; Miyazaki, T.; Matsuzaki, Y.; Ochiai, N. The effects of taurine administration in rat skeletal muscles on exercise. *J. Orthop. Sci.* **2003**, *8*, 415–419. [CrossRef]
21. Matsuzaki, Y.; Miyazaki, T.; Miyakawa, S.; Bouscarel, B.; Ikegami, T.; Tanaka, N. Decreased taurine concentration in skeletal muscles after exercise for various durations. *Med. Sci. Sports Exerc.* **2002**, *34*, 793–797. [CrossRef] [PubMed]

22. Miyazaki, T.; Honda, A.; Ikegami, T.; Iwamoto, J.; Monma, T.; Hirayama, T.; Saito, Y.; Yamashita, K.; Matsuzaki, Y. Simultaneous quantification of salivary 3-hydroxybutyrate, 3-hydroxyisobutyrate, 3-hydroxy-3-methylbutyrate, and 2-hydroxybutyrate as possible markers of amino acid and fatty acid catabolic pathways by LC-ESI-MS/MS. *Springer Plus* **2015**, *4*, 494. [CrossRef] [PubMed]
23. Robinson, A.M.; Williamson, D.H. Physiological roles of ketone bodies as substrates and signals in mammalian tissues. *Physiol. Rev.* **1980**, *60*, 143–187. [CrossRef] [PubMed]
24. Brevetti, G.; di Lisa, F.; Perna, S.; Menabo, R.; Barbato, R.; Martone, V.D.; Siliprandi, N. Carnitine-related alterations in patients with intermittent claudication: Indication for a focused carnitine therapy. *Circulation* **1996**, *93*, 1685–1689. [CrossRef]
25. Constantin-Teodosiu, D.; Cederblad, G.; Hultman, E. PDC activity and acetyl group accumulation in skeletal muscle during prolonged exercise. *J. Appl. Physiol.* **1992**, *73*, 2403–2407. [CrossRef]
26. Meienberg, F.; Loher, H.; Bucher, J.; Jenni, S.; Krusi, M.; Kreis, R.; Boesch, C.; Betz, M.J.; Christ, E. The effect of exercise on intramyocellular acetylcarnitine (AcCtn) concentration in adult growth hormone deficiency (GHD). *Sci. Rep.* **2019**, *9*, 19431. [CrossRef]
27. St Amand, T.A.; Spriet, L.L.; Jones, N.L.; Heigenhauser, G.J. Pyruvate overrides inhibition of PDH during exercise after a low-carbohydrate diet. *Am. J. Physiol. Endocrinol. Metab.* **2000**, *279*, E275–E283. [CrossRef] [PubMed]
28. Bremer, J. Carnitine–metabolism and functions. *Physiol. Rev.* **1983**, *63*, 1420–1480. [CrossRef] [PubMed]
29. Bezaire, V.; Bruce, C.R.; Heigenhauser, G.J.; Tandon, N.N.; Glatz, J.F.; Luiken, J.J.; Bonen, A.; Spriet, L.L. Identification of fatty acid translocase on human skeletal muscle mitochondrial membranes: Essential role in fatty acid oxidation. *Am. J. Physiol. Endocrinol. Metab.* **2006**, *290*, E509–E515. [CrossRef] [PubMed]
30. Miyazaki, T.; Nakamura, Y.; Ebina, K.; Mizushima, T.; Ra, S.G.; Ishikura, K.; Matsuzaki, Y.; Ohmori, H.; Honda, A. Increased N-acetyltaurine in the skeletal muscle after endurance exercise in rat. *Adv. Exp. Med. Biol.* **2017**, *975*, 403–411. [CrossRef]
31. Miyazaki, T.; Ishikura, K.; Honda, A.; Ra, S.G.; Komine, S.; Miyamoto, Y.; Ohmori, H.; Matsuzaki, Y. Increased N-acetyltaurine in serum and urine after endurance exercise in human. *Adv. Exp. Med. Biol.* **2015**, *803*, 53–62. [CrossRef] [PubMed]
32. Hoy, A.J.; Brandon, A.E.; Turner, N.; Watt, M.J.; Bruce, C.R.; Cooney, G.J.; Kraegen, E.W. Lipid and insulin infusion-induced skeletal muscle insulin resistance is likely due to metabolic feedback and not changes in IRS-1, Akt, or AS160 phosphorylation. *Am. J. Physiol. Endocrinol. Metab.* **2009**, *297*, E67–E75. [CrossRef] [PubMed]
33. Katz, A.; Raz, I.; Spencer, M.K.; Rising, R.; Mott, D.M. Hyperglycemia induces accumulation of glucose in human skeletal muscle. *Am. J. Physiol.* **1991**, *260*, R698–R703. [CrossRef] [PubMed]
34. Fueger, P.T.; Lee-Young, R.S.; Shearer, J.; Bracy, D.P.; Heikkinen, S.; Laakso, M.; Rottman, J.N.; Wasserman, D.H. Phosphorylation barriers to skeletal and cardiac muscle glucose uptakes in high-fat fed mice: Studies in mice with a 50% reduction of hexokinase II. *Diabetes* **2007**, *56*, 2476–2484. [CrossRef]
35. Furler, S.M.; Jenkins, A.B.; Storlien, L.H.; Kraegen, E.W. In vivo location of the rate-limiting step of hexose uptake in muscle and brain tissue of rats. *Am. J. Physiol.* **1991**, *261*, E337–E347. [CrossRef]
36. Furler, S.M.; Oakes, N.D.; Watkinson, A.L.; Kraegen, E.W. A high-fat diet influences insulin-stimulated posttransport muscle glucose metabolism in rats. *Metabolism* **1997**, *46*, 1101–1106. [CrossRef]
37. Miyamoto, Y.; Miyazaki, T.; Honda, A.; Shimohata, H.; Hirayama, K.; Kobayashi, M. Retention of acetylcarnitine in chronic kidney disease causes insulin resistance in skeletal muscle. *J. Clin. Biochem. Nutr.* **2016**, *59*, 199–206. [CrossRef] [PubMed]
38. Johnson, C.H.; Patterson, A.D.; Krausz, K.W.; Lanz, C.; Kang, D.W.; Luecke, H.; Gonzalez, F.J.; Idle, J.R. Radiation metabolomics. 4. UPLC-ESI-QTOFMS-Based metabolomics for urinary biomarker discovery in gamma-irradiated rats. *Radiat. Res.* **2011**, *175*, 473–484. [CrossRef]
39. Miyazaki, T.; Honda, A.; Ikegami, T.; Matsuzaki, Y. The role of taurine on skeletal muscle cell differentiation. *Adv. Exp. Med. Biol.* **2013**, *776*, 321–328. [CrossRef] [PubMed]
40. Namikawa-Kanai, H.; Miyazaki, T.; Matsubara, T.; Shigefuku, S.; Ono, S.; Nakajima, E.; Morishita, Y.; Honda, A.; Furukawa, K.; Ikeda, N. Comparison of the amino acid profile between the nontumor and tumor regions in patients with lung cancer. *Am. J. Cancer Res.* **2020**, *10*, 2145–2159. [PubMed]
41. Honda, A.; Salen, G.; Matsuzaki, Y.; Batta, A.K.; Xu, G.; Leitersdorf, E.; Tint, G.S.; Erickson, S.K.; Tanaka, N.; Shefer, S. Differences in hepatic levels of intermediates in bile acid biosynthesis between Cyp27(-/-) mice and CTX. *J. Lipid Res.* **2001**, *42*, 291–300. [CrossRef]

Article

N-Chlorotaurine Reduces the Lung and Systemic Inflammation in LPS-Induced Pneumonia in High Fat Diet-Induced Obese Mice

Nguyen Khanh Hoang [1], Eiji Maegawa [1], Shigeru Murakami [1], Stephen W. Schaffer [2] and Takashi Ito [1,*]

[1] Faculty of Bioscience and Biotechnology, Fukui Prefectural University, Eiheiji 910-1195, Japan; s2073003@g.fpu.ac.jp (N.K.H.); s1821045@g.fpu.ac.jp (E.M.); murakami@fpu.ac.jp (S.M.)

[2] Department of Pharmacology, College of Medicine, University of South Alabama, Mobile, AL 36688, USA; sschaffe@southalabama.edu

* Correspondence: tito@fpu.ac.jp; Tel.: +81-776-61-6000

Abstract: Lung infection can evoke pulmonary and systemic inflammation, which is associated with systemic severe symptoms, such as skeletal muscle wasting. While *N*-chlorotaurine (also known as taurine chloramine; TauCl) has anti-inflammatory effects in cells, its effects against pulmonary and systemic inflammation after lung infection has not been elucidated. In the present study, we evaluated the anti-inflammatory effect of the taurine derivative, TauCl against *Escherichia coli*-derived lipopolysaccharide (LPS)-induced pneumonia in obese mice maintained on a high fat diet. In this study, TauCl was injected intraperitoneally 1 h before intratracheal LPS administration. While body weight was decreased by 7.5% after LPS administration, TauCl treatment suppressed body weight loss. TauCl also attenuated the increase in lung weight due to lung edema. While LPS-induced acute pneumonia caused an increase in cytokine/chemokine mRNA expression, including that of IL-1β, -6, TNF-α, MCP-1, TauCl treatment attenuated IL-6, and TNF-alpha expression, but not IL-1β and MCP-1. TauCl treatment partly attenuated the elevation of the serum cytokines. Furthermore, TauCl treatment alleviated skeletal muscle wasting. Importantly, LPS-induced expression of Atrogin-1, MuRF1 and IκB, direct or indirect targets for NFκB, were suppressed by TauCl treatment. These findings suggest that intraperitoneal TauCl treatment attenuates acute pneumonia-related pulmonary and systemic inflammation, including muscle wasting, in vivo.

Keywords: taurine; *N*-chlorotaurine; lung inflammation; cytokine; muscle wasting; atrophy

Citation: Khanh Hoang, N.; Maegawa, E.; Murakami, S.; Schaffer, S.W.; Ito, T. *N*-Chlorotaurine Reduces the Lung and Systemic Inflammation in LPS-Induced Pneumonia in High Fat Diet-Induced Obese Mice. *Metabolites* **2022**, *12*, 349. https://doi.org/10.3390/metabo12040349

Academic Editor: Markus R. Meyer

Received: 11 March 2022
Accepted: 12 April 2022
Published: 14 April 2022

1. Introduction

When an organism is invaded by a pathogen such as bacteria and virus, the innate immune system, including neutrophils and macrophages, is activated to secrete antimicrobial substances and phagocytose the pathogen. Cytokines are secreted to activate the surrounding immune system, eliminate the pathogen and protect the organism. However, a strong immune response can lead to excessive production of cytokines not only from immune cells but also from non-immune cells, such as fibroblasts and vascular endothelial cells, a phenomenon known as a cytokine storm [1]. Cytokine storms can be triggered not only by pathogens, but by cancer, autoimmune diseases, and immunotherapies, causing a lethal systemic immune response due to elevated circulating cytokines. Thus, the activated immune system attacks its own tissues rather than pathogens, resulting in multiple organ failure.

Taurine is a beta-amino acid found in high concentrations in mammalian tissues [2,3]. Taurine exerts a variety of biological actions, including antioxidation, modulation of ion movement, osmoregulation, and stabilization of macromolecules etc. [4,5] Moreover, treatment of taurine benefits many kinds of pathologies [6,7]. Concerning lung inflammation, previous studies have demonstrated that oral taurine treatment reduces lung inflammation in bleomycin-induced lung injury [8].

Taurine is present in immune system cells, such as macrophages and neutrophils, at levels as high as 35 mM [9]. When an organism is infected by pathogens, immune cells produce hypochlorous acid to kill the pathogens, while taurine reacts with excessive hypochlorous acid to produce *N*-chlorotaurine (also known as taurine chloramine; TauCl), which reduces the high levels of hypochlorous acid and thus its toxicity to surrounding host cells. TauCl also has a broad spectrum of pathogenic microbial (viral, fungal, and bacterial) killing effects [10–13] and has been reported to inhibit excessive inflammatory reactions through suppression of the production of inflammatory cytokines [14]. In in vitro experiments, TauCl reduces the expression of prostaglandin E2, nitric oxide (NO), and inflammatory cytokines, such as IL-1, -6, -8 and TNF-α [14]. These effects on the regulation of inflammatory cytokine expression have been confirmed not only in immune cells but also in adipocytes and synovial fibroblasts of bone tissue [15,16]. The mechanism underlying the anti-inflammatory action of TauCl is partly associated with the modulation of nuclear factor-κB (NF-κB) [17]. While these results suggest that TauCl may be useful against the excessive production of cytokines that are thought to be involved in the severity of pneumonia, including bacterial infection disease, there is little information about the in vivo effects of TauCl on excessive inflammation.

Lipopolysaccharide (LPS) is a component of the outer membrane of Gram-negative bacteria, which can induce inflammatory responses and is commonly used to study pulmonary inflammation and acute respiratory disease in mice [18]. It binds to cellular Toll-like receptor 4 (TLR4) and induces cytokine expression through NF-κB activation. In the present study, we investigated the effect of TauCl against LPS-induced acute pulmonary/systemic inflammation in obese mice. Moreover, we also evaluated its effect against skeletal muscle atrophy, one of the important systemic reactions of LPS.

2. Results

2.1. Effects on Lung Inflammation

In our preliminary study, we found that mice maintained on a high fat diet to induce obesity showed more severe weight loss and lung inflammation after LPS administration than normal young mice. Indeed, normal young mice, but not high fat diet-induced obese mice, may start to recover within 2 days after intratracheal LPS administration (Nguyen et al. 2022 in press). Therefore, in the present study, LPS was administered to obese mice, which were fed a high fat diet for 10 weeks to induce obesity and lung inflammation; in some animals the effect of TauCl was examined. Weight loss was significantly suppressed by the administration of TauCl (Figure 1A). At autopsy 2 days later, LPS administration had resulted in a significant increase in lung weight, confirming the site of inflammation (Figure 1B,C). TauCl treatment slowed the inflammatory response in lung tissue, with the increase in lung weight being predominantly suppressed, indicating that edema and immune cell infiltration in the lung are diminished by TauCl treatment. However, TauCl minimized the decrease in spleen weight induced by LPS.

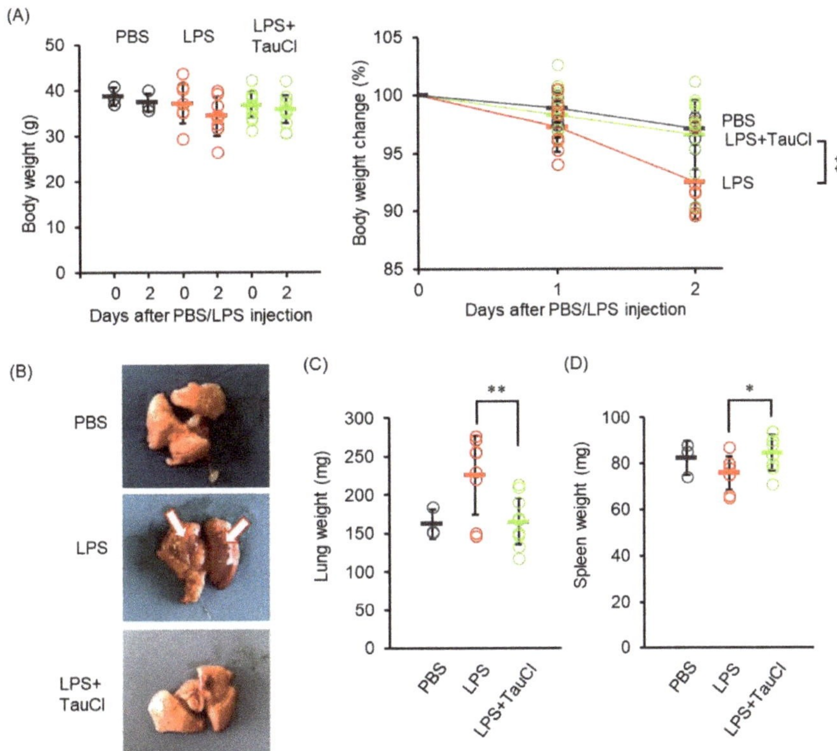

Figure 1. Effect of TauCl on lung inflammation induced by intratracheal LPS. (**A**) The changes in body weight after intratracheal PBS or LPS injection with TauCl. Body weight was monitored after LPS injection. (**B**) Representative lung images isolated from mice at 2 days after LPS injection. White arrows indicate the inflammation site. (**C**,**D**) The weight of lungs (**C**) and spleens (**D**) was shown. TauCl was intraperitoneally pretreated 1 h prior to LPS injection in the TauCl-treated group (LPS + TauCl). Values shown represent means ± SD. n = 3 (PBS), 9 (LPS), and 11 (LPS + TauCl). * $p < 0.05$, ** $p < 0.01$ (Tukey-Kramer's t-test).

2.2. Effect of TauCl on the LPS-Induced Cytokine mRNA Expression in Lung

To explore the effect of TauCl on cytokine expression induced by intratracheal LPS administration, RNA purified from lungs of mice treated with LPS with or without TauCl were analyzed. As shown in Figure 2, cytokines were elevated after LPS injection, but TauCl predominantly decreased the expression of IL-6 and TNF-α (Figure 2). Meanwhile the gene levels of IL-1β and inflammasomes were upregulated by LPS injection but not suppressed by TauCl.

2.3. Effects of TauCl on Systemic Cytokine Level in LPS-Treated Mice

To determine the effect of TauCl on the circulation of cytokines of pneumonia model animals, serum levels of cytokines, including, IL-6, and TNF-α, were measured (Figure 3). TauCl significantly suppressed the increase in serum TNF-α levels, while tending to lower IL-6 levels although the effect was not statistically significant.

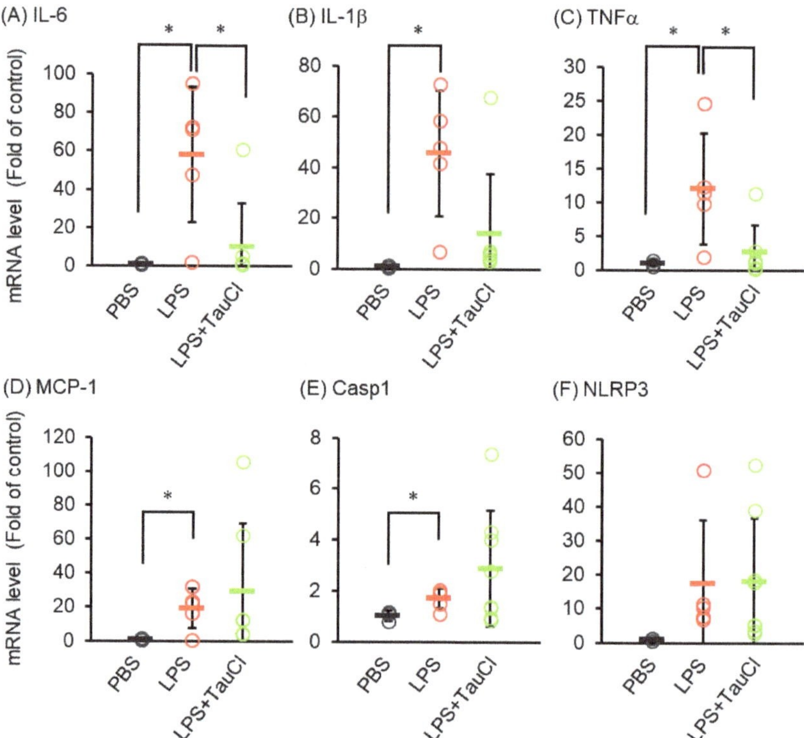

Figure 2. Effect of TauCl on LPS-induced expression of cytokine/chemokine mRNA in the lung. The lungs were isolated from mice 2 days after PBS or LPS injection. TauCl was intraperitoneally pretreated 1 h prior to LPS injection (LPS + TauCl). The expression of IL-6 (**A**), IL-1β (**B**), TNFα (**C**), MCP-1 (**D**), Casp1 (**E**) and NLRP3 (**F**) were measured by qPCR. The expression level was normalized by the expression level of GAPDH. Values are shown fold of control (PBS). $n = 3$ (PBS), 8 (LPS), and 11 (LPS + TauCl). * $p < 0.05$ (Tukey–Kramer's *t*-test for IL-6, IL-1β, MCP-1, Casp1 and Mann-Whitney U-test for TNFα).

2.4. Effects of TauCl against Muscle Wasting

Elevated levels of circulating cytokines cause several systemic reactions, including muscle wasting [19]. As shown in Figure 1A, TauCl attenuates the loss of body weight after intratracheal LPS administration, implying that TauCl alleviates muscle wasting. We further investigated whether TauCl treatment contributes to the inhibition of muscle wasting in tibialis anterior muscle. Histological examination showed that myofiber size was decreased within 2 days of LPS administration and that TauCl significantly suppressed the decrease in cross sectional area of the fibers (Figure 4). In addition, the expression of Atrogin-1 and MurF1, which are markers of skeletal muscle atrophy, increase in response to LPS administration. By suppressing their expression (Figure 5A,B), TauCl alleviates skeletal muscle wasting induced by intratracheal LPS injection.

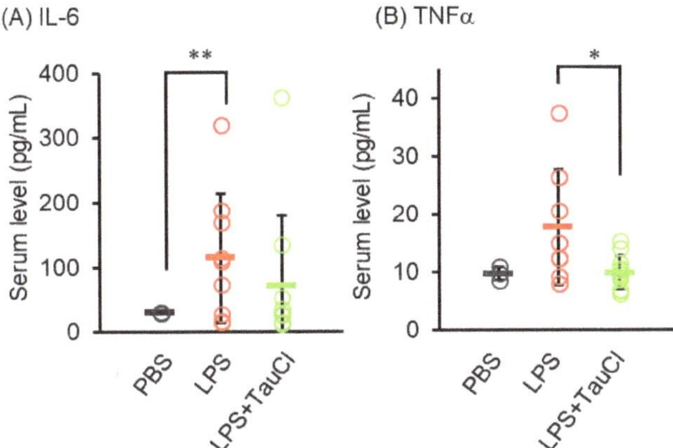

Figure 3. Effect of TauCl and taurine on serum cytokine level. Blood was collected from mice 2 days after PBS or LPS injection. Serum IL-6 (**A**) and TNF-α (**B**) levels were measured by ELISA. Values shown represent means ± SD. $n = 3$ (PBS), 8 (LPS), and 11 (LPS + TauCl). * $p < 0.05$, ** $p < 0.01$ (Tukey–Kramer's t-test).

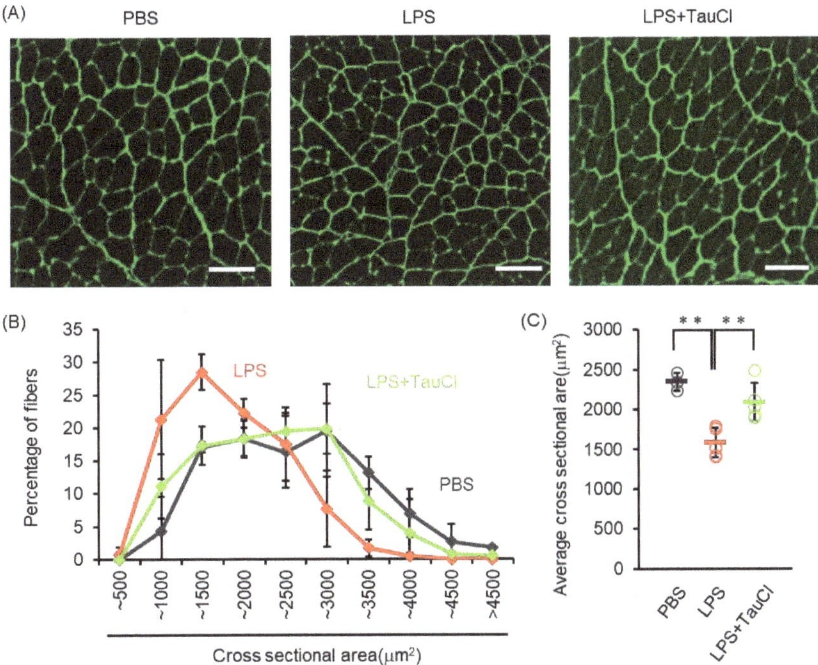

Figure 4. Effect of TauCl on skeletal muscle wasting induced by intratracheal LPS. The skeletal muscles (tibialis anterior muscles) were isolated from mice 2 days after PBS or LPS injection. TauCl was intraperitoneally pretreated 1 h prior to LPS injection (LPS + TauCl). Frozen section of skeletal muscles were immunostained with anti-laminin antibody. (**A**) Representative cross sectional image of skeletal muscle. Scale bars indicate 100 μm. (**B**) Cross sectional area of skeletal muscle fiber was measured from the immunostained images. Values indicate the frequency (percentage) of each size of muscle fiber. (**C**) The average of cross sectional area of skeletal muscle of each mouse was shown. $n = 3$ (PBS), 5 (LPS), and 5 (LPS + TauCl). ** $p < 0.01$ (Tukey–Kramer's t-test).

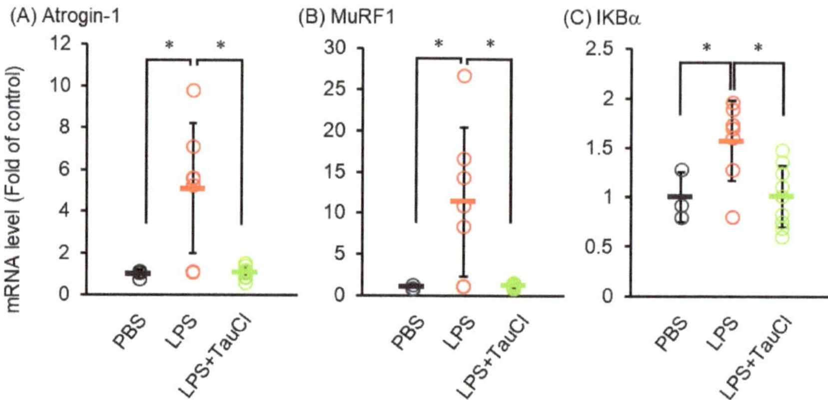

Figure 5. Effect of TauCl on the atrogenes and IKBa expression in skeletal muscle in LPS-treated mice. The tibialis anterior muscles were isolated from mice 2 days after PBS or LPS injection. TauCl was intraperitoneally pretreated 1 h prior to LPS injection (LPS + TauCl). The expression of Atrogin-1 (**A**), MurF1 (**B**) and IkBα (**C**) were measured by qPCR. The expression level was normalized by the expression level of GAPDH. Values are shown fold of control (PBS). n = 3 (PBS), 7 (LPS), and 7 (LPS + TauCl). * $p < 0.05$ (Tukey–Kramer's t-test).

Finally, to investigate whether NF-κB is involved in the suppression of muscle atrophy by TauCl, the mRNA expression of IκBα, a product of NF-κB, was analyzed in muscles from LPS-injected mice treated with or without TauCl. While intratracheal LPS injection increased IκBα mRNA of skeletal muscle, TauCl treatment suppressed them (Figure 5C).

3. Discussion

While accumulating evidence from in vitro studies suggest that TauCl alleviates inflammation in many types of inflammatory cells, there have been no studies on pulmonary inflammation and the systemic cytokine storm in animals. In the present study, we investigated the effects of intraperitoneal administration of TauCl to obese mice that had previously been treated with LPS to produce a pneumonia-like condition. TauCl treatment suppressed LPS-induced lung edema and attenuated the production of pro-inflammatory cytokines in the lung of those animals, indicating that TauCl suppressed the excessive immune response in lungs. In addition, systemic cytokine elevation induced by intratracheal injection of LPS was partially diminished. These results suggest that TauCl has an inhibitory effect on the severity of inflammation induced by infection-induced pneumonia. In addition, TauCl suppressed the decline in spleen weight, indicating that TauCl may affect the distribution and function of immune cells. Further studies are required to determine how TauCl regulates immune cells in their inflammatory state.

An increase in systemic cytokines causes a variety of systematic symptoms. In the present study, we demonstrated that TauCl treatment alleviated muscle wasting in the LPS-induced lung inflammation model. This anti-wasting role may be caused in part by suppression of circulating inflammatory cytokines. Both IL-6 and TNFα can activate the catabolic signaling pathway in skeletal muscle leading to muscle wasting [20]. Previous in vitro study revealed that TauCl inhibits the cellular response against cytokines, including TNFα and IL-1β [14], implying that TauCl directly prevents the action of cytokines in skeletal muscle. In contrast, a recent study reported that LPS administered according to the intratracheal method can migrate into the blood [21]. Hence, it is possible that TauCl blocks the systemic actions of LPS. Other potential mechanisms may be involved in the anti-wasting actions of TauCl in this model. In the present study, we observed that TauCl suppressed the expression of atrogenes, such as Atrogin-1 and MuRF1, as well as reduced the levels of IκB. MuRF1 and IκBα are the direct downstream targets of NF-κB [20,22] while Atrogin-1 is regulated by NF-κB via nuclear phosphatase SCP4, which is another target of NF-κB [23]. Therefore, these observations suggest a direct action of TauCl on skeletal muscle through the inhibition of NF-κB, which may partly involve an anti-muscle wasting effect.

It has previously been shown that young (8 to 10-week-old) mice administered LPS lose weight and their lungs accumulate fluid (Nguyen et al. 2022 in press). However, weight loss and lung edema were more severe in obese mice. Since adipose tissue also produces cytokines and is the cause of chronic inflammation in obesity, it is possible that LPS administration amplifies the inflammatory response. In this regard it is possible that TauCl can inhibit cytokine production in macrophages in adipose tissue [24]. The results in this study need to be further studied, as TauCl might not only suppress excessive cytokine production by acting on immune cells, but also affect the production of cytokines by adipose tissue macrophage.

In the present study, administration of TauCl was intraperitoneal route, since TauCl is considered to be unstable due to the elimination by anti-oxidation mechanism, such as glutathione (GSH) and ascorbate, or the other amino acids [9,14,25,26]. Yet, we observed that some acute inflammatory events were partially suppressed by TauCl; indeed, some of the injected TauCl may have been transferred to the tissues or may have been dechlorinated to taurine by an anti-oxidation mechanism in body fluid. Since the method of measuring TauCl from crude samples, such as blood and tissues, has not been established, we could not clarify the pharmacokinetics of TauCl in the body. An important limitation of this study is that our research could not clarify how much TauCl itself is actioned in cells. Taurine, which were converted from TauCl, might partly contribute to reduce lung or systemic inflammation. Indeed, it has been reported that systemic (intraperitoneal or intravenous) administration of taurine alleviated inflammatory responses in vivo studies [27,28].

Kwaśny-Krochin et al. reported that a single dose of TauCl in collagen-induced arthritis delayed the onset of inflammation, but did not diminish the severity of arthritis and the onset of cytokine production at the end of experiments [29]. Meanwhile, Wang et al. observed that continuous subcutaneous injections of TauCl attenuated arthritic symptoms and synovial inflammation in collagen-induced arthritic mice [30]. Thus, it is logical to conclude that sustained administration of TauCl may be more effective than a single administration of TauCl in suppressing the production of pro-inflammatory cytokines in the lung of LPS-treated mice with acute lung inflammation.

A novel type of coronavirus infection (COVID-19) has been spreading around the world since 2020, and by the end of 2021, about 5.4 million people have lost their lives (Available online: https://covid19.who.int/ (accessed on 29 December 2021). In patients with COVID-19, there is a correlation between the levels of cytokines in the blood and the severity of the disease, suggesting that excessive activation of the immune response contributes to the severity of the disease [31]. Therefore, it is desirable to develop therapies to limit the immune response, as well as antiviral drugs against the causative virus. Recent studies have revealed that TauCl can inactivate SARS-CoV-2 in vitro [32]. In addition to its antimicrobial role, the present findings raise the possibility that TauCl treatment might alleviate the severe symptoms associated with acute pneumonia, including COVID-19.

4. Materials and Methods

4.1. TauCl Synthesis

TauCl was synthesized as previously described [29]. In brief, 20 mM sodium hypochlorite solution (Nacalai tesque, Kyoto, Japan) was added to 24 mM taurine (Nacalai tesque) dissolved in 50 mM phosphate buffer (pH 7.4). Production of TauCl was spectrophotometrically confirmed by UV absorption spectra (220–340 nm, Amax = 252 nm). The concentration of TauCl was determined using the molar extinction coefficient (ε252 nm = 415 M^{-1} cm^{-1}) [17].

4.2. Animal Care

All experimental procedures were approved by the Institutional Animal Care and Use Committee of the Fukui Prefectural University. Male C57BL/6J mice (6-week-old, Japan Crea) were used for this study. Mice were fed 60% fat-containing chow (Oriental Yeast, Tokyo, Japan) for 10 weeks to induce obesity [33]. The mice had access to water ad libitum, and were maintained on a 12-h light/dark cycle.

4.3. LPS-Induced Pneumonia Model

Pneumonia was induced by intratracheal injection of a LPS solution (O-111: 7 mg/kg body weight, Fuji Film Co. Ltd., Tokyo, Japan) under anesthesia with a combination of medetomidine hydrochloride (0.3 mg/kg, Nippon Zenyaku Kogyo Co., Ltd., Fukushima, Japan), midazolam (4.0 mg/kg, Astellas Pharma Inc., Tokyo, Japan) and butorphanol (5.0 mg/kg, Meiji Seika Pharma Co., Ltd., Tokyo, Japan), as previously described [34]. An equivalent amount of vehicle (PBS) was injected for the control mice. After injection, atipamezole (0.3 mg/kg, Nippon Zenyaku Kogyo Co., Ltd.) was treated to recover from anesthesia. One hour before LPS injection, the mice were intraperitoneally administered a TauCl solution (5 mM, 250 μL) or PBS. Two days after LPS injection, mice were euthanized, and whole blood was immediately collected by retro-orbital bleeding. Then, tissues were collected, immediately frozen in liquid nitrogen and stored at −80 °C until use.

4.4. mRNA Measurement

Total RNA was extracted from mouse tissues by using Sepasol (Nacalai tesque) according to the manufacturer's protocol. cDNA was generated from total RNA by reverse transcription with Rever Tra Ace (Toyobo, Japan). Quantitative RT-PCR analyses were performed by using qTOWER^3 (Analytik Jena GmBH, Jena, Germany) with KAPA SYBR Fast qPCR Kit (Kapa Biosystems, Inc., Wilmington, MA, USA). The sequence for primers were listed in Table 1. Some of the primers were previously described [35] (Khanh2022 in press). GAPDH is used as an internal control. Data were analyzed using the ΔΔCt method.

Table 1. Primers for quantitative PCR.

Gene	Forward Primer (5′→3′)	Reverse Primer (5′→3′)
IL-1β	CCTTG GGCCT CAAAG GAAAG A	TTGCT TGGGA TCCAC ACTCT CC
IL-6	CACTT CACAA GTCGG AGGCT T	GAATT GCCAT TGCAC AACTC TTTTC
TNF-α	CAAAA TTCGA GTGAC AAGCC TGTA	CACCA CTAGT TGGTT GTCTT TGAGA
MCP-1	CTGTC ATGCT TCTGG GCCTG	GGCGT TAACT GCATC TGGCT GA
Casp1	GCATG CCGTG GAGAG AAACA	ATGGG CCTTC TTAAT GCCAT CAT
NLRP3	GCAGA GCCTA CAGTT GGGTG	CTTCC ACGCC TACCA GGAAA T
Atrogin-1	TTCAG CAGCC TGAAC TACGA	AGTAT CCATG GCGCT CCTTC
MuRF1	GCGTG ACCAC AGAGG GTAAA	CTCTG CGTCC AGAGC GTG
IκBα	TAGCA GTCTT GACGC AGACC	GACAC GTGTG GCCAT TGTAG
GAPDH	GCCGG TGCTG AGTAT GTCGT	CCCTT TTGGC TCCAC CCTT

4.5. Plasma Cytokine Assay

Serum cytokine concentrations (IL-6 and TNF-α) were measured by ELISA by using the Mouse IL-6 Quantikine ELISA Kit (R&D Systems, Inc., Abingdon, UK), Mouse TNF-αELISA Kit (Fujifilm, Osaka, Japan) respectively, according to the manufacturer's protocol.

4.6. Tissue Section and Immunostaining

Sections from frozen tissues were cut by cryostat (Leica Microsystems, Wetzlar, Germany). For detection of Laminin, the frozen section was immunostained by using anti-laminin antibody (ab80, Abcam, Cambridge, UK; 1:100) and Alexa Fluor 488-conjugated second antibody (Life Technologies; 1:400) with Can Get Signal Immunostain according to manufacturer's protocol (Toyobo, Osaka, Osaka). Images were acquired with microscopes (BZ-9000, Keyence, Osaka, Osaka) equipped with imaging software (BZ-II, Keyence). Cross-sectional area was calculated by using ImageJ software (version: 1.53e).

4.7. Statistical Analysis

Data are presented as the means ± standard deviation. Microsoft Excel and R software packages were used for statistical analysis of the data. Data were tested for normal distribution using the Shapiro–Wilk normality test. The non-parametric Mann–Whitney U-test and the parametric independent *t*-test and Tukey-Kramer's *t*-test were used to determine statistical significance between groups. The Grubbs test was used to detect and reject outliers. Differences were considered statistically significant when the calculated *p*-value was less than 0.05.

Author Contributions: Conceptualization, T.I.; methodology, N.K.H. and T.I.; validation, N.K.H., E.M. and T.I.; formal analysis, T.I.; investigation, N.K.H., E.M. and T.I.; resources, T.I.; data curation, T.I.; writing—original draft preparation, N.K.H. and T.I.; writing—review and editing, S.M., S.W.S. and T.I.; visualization, T.I.; supervision, T.I.; project administration, T.I.; funding acquisition, T.I. All authors have read and agreed to the published version of the manuscript.

Funding: This research was funded by Step-up Research Promotion Grant from Fukui Prefectural University, and the grant for Scientific Research from Lotte Research Promotion Grant (to Takashi Ito).

Institutional Review Board Statement: The animal study protocol was approved by the Institutional Review Board of Fukui Prefectural University (protocol code 2020-M10-2 (22 June 2020)).

Informed Consent Statement: Not applicable.

Data Availability Statement: Data is contained within the article.

Conflicts of Interest: The authors declare no conflict of interest.

References

1. Fajgenbaum, D.C.; June, C.H. Cytokine Storm. *N. Engl. J. Med.* **2020**, *383*, 2255–2273. [CrossRef] [PubMed]
2. Chesney, R.W. Taurine: Its Biological Role and Clinical Implications. *Adv. Pediatr.* **1985**, *32*, 1–42. [PubMed]
3. Ito, T.; Kimura, Y.; Uozumi, Y.; Takai, M.; Muraoka, S.; Matsuda, T.; Ueki, K.; Yoshiyama, M.; Ikawa, M.; Okabe, M.; et al. Taurine depletion caused by knocking out the taurine transporter gene leads to cardiomyopathy with cardiac atrophy. *J. Mol. Cell. Cardiol.* **2008**, *44*, 927–937. [CrossRef] [PubMed]
4. Tsubotani, K.; Maeyama, S.; Murakami, S.; Schaffer, S.W.; Ito, T. Taurine suppresses liquid–liquid phase separation of lysozyme protein. *Amino Acids* **2021**, *53*, 745–751. [CrossRef]
5. Bhat, M.A.; Ahmad, K.; Khan, M.S.A.; Bhat, M.A.; Almatroudi, A.; Rahman, S.; Jan, A.T. Expedition into Taurine Biology: Structural Insights and Therapeutic Perspective of Taurine in Neurodegenerative Diseases. *Biomolecules* **2020**, *10*, 863. [CrossRef]
6. Ito, T.; Schaffer, S.; Azuma, J. The effect of taurine on chronic heart failure: Actions of taurine against catecholamine and angiotensin II. *Amino Acids* **2014**, *46*, 111–119. [CrossRef]
7. Ito, T.; Schaffer, S.W.; Azuma, J. The potential usefulness of taurine on diabetes mellitus and its complications. *Amino Acids* **2011**, *42*, 1529–1539. [CrossRef]
8. Schuller-Levis, G.B.; Gordon, R.E.; Wang, C.; Park, E. Taurine Reduces Lung Inflammation and Fibrosis Caused by Bleomycin. *Adv. Exp. Med. Biol.* **2003**, *526*, 395–402. [CrossRef]
9. Marcinkiewicz, J.; Kontny, E. Taurine and inflammatory diseases. *Amino Acids* **2014**, *46*, 7–20. [CrossRef]
10. Dudani, A.; Martyres, A.; Fliss, H. Short Communication:Rapid Preparation of Preventive and Therapeutic Whole-Killed Retroviral Vaccines Using the Microbicide Taurine Chloramine. *AIDS Res. Hum. Retrovir.* **2008**, *24*, 635–642. [CrossRef]
11. Nagl, M.; Larcher, C.; Gottardi, W. Activity of N-chlorotaurine against herpes simplex- and adenoviruses. *Antivir. Res.* **1998**, *38*, 25–30. [CrossRef]
12. Nagl, M.; Arnitz, R.; Lackner, M. N-Chlorotaurine, a Promising Future Candidate for Topical Therapy of Fungal Infections. *Mycopathologia* **2018**, *183*, 161–170. [CrossRef] [PubMed]
13. Gottardi, W.; Nagl, M. N-chlorotaurine, a natural antiseptic with outstanding tolerability. *J. Antimicrob. Chemother.* **2010**, *65*, 399–409. [CrossRef] [PubMed]
14. Kim, C.; Cha, Y.-N. Taurine chloramine produced from taurine under inflammation provides anti-inflammatory and cytoprotective effects. *Amino Acids* **2013**, *46*, 89–100. [CrossRef] [PubMed]
15. Kim, K.S.; Ji, H.-I.; Chung, H.; Kim, C.; Lee, S.H.; Lee, Y.-A.; Yang, H.-I.; Yoo, M.C.; Hong, S.J. Taurine chloramine modulates the expression of adipokines through inhibition of the STAT-3 signaling pathway in differentiated human adipocytes. *Amino Acids* **2013**, *45*, 1415–1422. [CrossRef]

16. Kontny, E.; Grabowska, A.; Kowalczewski, J.; Kurowska, M.; Ku-Rowska, M.; Janicka, I. Taurine Chloramine Inhibi-tion of Cell Proliferation and Cytokine Production by Rheumatoid Arthritis Fibroblast-Like Syno-Viocytes. *Arthritis Rheum. Off. J. Am. Coll. Rheumatol.* **1999**, *42*, 2552–2560. [CrossRef]
17. Kanayama, A.; Inoue, J.-I.; Sugita-Konishi, Y.; Shimizu, M.; Miyamoto, Y. Oxidation of IκBα at Methionine 45 Is One Cause of Taurine Chloramine-induced Inhibition of NF-κB Activation. *J. Biol. Chem.* **2002**, *277*, 24049–24056. [CrossRef]
18. Domscheit, H.; Hegeman, M.A.; Carvalho, N.; Spieth, P.M. Molecular Dynamics of Lipopolysaccharide-Induced Lung Injury in Rodents. *Front. Physiol.* **2020**, *11*, 36. [CrossRef]
19. Webster, J.M.; Kempen, L.J.A.P.; Hardy, R.S.; Langen, R.C.J. Inflammation and Skeletal Muscle Wasting During Cachexia. *Front. Physiol.* **2020**, *11*, 1449. [CrossRef]
20. McKinnell, I.W.; Rudnicki, M.A. Molecular Mechanisms of Muscle Atrophy. *Cell* **2004**, *119*, 907–910. [CrossRef]
21. Bivona, J.J.; Crymble, H.M.; Guigni, B.A.; Stapleton, R.D.; Files, D.C.; Toth, M.J.; Poynter, M.E.; Suratt, B.T. Macrophages augment the skeletal muscle proinflammatory response through TNFα following LPS-induced acute lung injury. *FASEB J.* **2021**, *35*, e21462. [CrossRef] [PubMed]
22. Ruland, J. Return to homeostasis: Downregulation of NF-κB responses. *Nat. Immunol.* **2011**, *12*, 709–714. [CrossRef] [PubMed]
23. Cheung, W.W.; Hao, S.; Mak, R.H. Emerging roles of nuclear phosphatase SCP4 in CKD-associated muscle wasting. *Kidney Int.* **2017**, *92*, 281–283. [CrossRef] [PubMed]
24. Lin, S.; Hirai, S.; Yamaguchi, Y.; Goto, T.; Takahashi, N.; Tani, F.; Mutoh, C.; Sakurai, T.; Murakami, S.; Yu, R.; et al. Taurine improves obesity-induced inflammatory responses and modulates the unbalanced phenotype of adipose tissue macrophages. *Mol. Nutr. Food Res.* **2013**, *57*, 2155–2165. [CrossRef] [PubMed]
25. Peskin, A.V.; Winterbourn, C.C. Kinetics of the reactions of hypochlorous acid and amino acid chloramines with thiols, methionine, and ascorbate. *Free Radic. Biol. Med.* **2001**, *30*, 572–579. [CrossRef]
26. Peskina, A.V.; Midwinter, R.G.; Harwood, D.T.; Winterbourn, C.C. Chlorine transfer between glycine, taurine, and histamine: Reaction rates and impact on cellular reactivity. *Free Radic. Biol. Med.* **2005**, *38*, 397–405. [CrossRef]
27. Nakajima, Y.; Osuka, K.; Seki, Y.; Gupta, R.C.; Hara, M.; Takayasu, M.; Wakabayashi, T. Taurine Reduces Inflammatory Responses after Spinal Cord Injury. *J. Neurotrauma* **2010**, *27*, 403–410. [CrossRef]
28. Liu, Y.; Li, F.; Zhang, L.; Wu, J.; Wang, Y.; Yu, H. Taurine alleviates lipopolysaccharide-induced liver injury by anti-inflammation and antioxidants in rats. *Mol. Med. Rep.* **2017**, *16*, 6512–6517. [CrossRef]
29. Kwaśny-Krochin, B.; Bobek, M.; Kontny, E.; Gluszko, P.; Biedroń, R.; Chain, B.; Maśliński, W.; Marcinkiewicz, J. Effect of taurine chloramine, the product of activated neutrophils, on the development of collagen-induced arthritis in DBA 1/J mice. *Amino Acids* **2002**, *23*, 419–426. [CrossRef]
30. Wang, Y.; Cha, Y.-N.; Kim, K.S.; Kim, C. Taurine chloramine inhibits osteoclastogenesis and splenic lymphocyte proliferation in mice with collagen-induced arthritis. *Eur. J. Pharmacol.* **2011**, *668*, 325–330. [CrossRef]
31. Cao, X. COVID-19: Immunopathology and its implications for therapy. *Nat. Rev. Immunol.* **2020**, *20*, 269–270. [CrossRef] [PubMed]
32. Lackner, M.; Rössler, A.; Volland, A.; Stadtmüller, M.; Müllauer, B.; Banki, Z.; Ströhle, J.; Luttick, A.; Fenner, J.; Stoiber, H.; et al. N-Chlorotaurine, a Novel Inhaled Virucidal Antiseptic Is Highly Active against Respiratory Viruses Including SARS-CoV-2 (COVID-19). 2020. Available online: https://assets.researchsquare.com/files/rs-118665/v1_covered.pdf?c=1631849287 (accessed on 12 April 2022). [CrossRef]
33. Ito, T.; Yoshikawa, N.; Ito, H.; Schaffer, S.W. Impact of taurine depletion on glucose control and insulin secretion in mice. *J. Pharmacol. Sci.* **2015**, *129*, 59–64. [CrossRef] [PubMed]
34. Langen, R.C.J.; Haegens, A.; Vernooy, J.H.J.; Wouters, E.F.M.; de Winther, M.; Carlsen, H.; Steele, C.; Shoelson, S.E.; Schols, A.M.W.J. NF-κB Activation Is Required for the Transition of Pulmonary Inflammation to Muscle Atrophy. *Am. J. Respir. Cell Mol. Biol.* **2012**, *47*, 288–297. [CrossRef] [PubMed]
35. Nguyen, K.H.; Ito, S.; Maeyama, S.; Schaffer, S.W.; Murakami, S.; Ito, T. In Vivo and In Vitro Study of *N*-Methyltaurine on Pharmacokinetics and Antimuscle Atrophic Effects in Mice. *ACS Omega* **2020**, *5*, 11241–11246. [CrossRef] [PubMed]

MDPI
St. Alban-Anlage 66
4052 Basel
Switzerland
Tel. +41 61 683 77 34
Fax +41 61 302 89 18
www.mdpi.com

Metabolites Editorial Office
E-mail: metabolites@mdpi.com
www.mdpi.com/journal/metabolites

www.ingramcontent.com/pod-product-compliance
Lightning Source LLC
LaVergne TN
LVHW070841170726
843515LV00005B/540

* 9 7 8 3 0 3 6 5 6 8 7 5 1 *